THE COMMERCIAL SPACE STATION

METHODS AND MARKETS

BY

ANDREW M. THORPE

authorHOUSE®

AuthorHouse™
1663 Liberty Drive, Suite 200
Bloomington, IN 47403
www.authorhouse.com
Phone: 1-800-839-8640

First published by AuthorHouse 12/8/2009

ISBN: 978-1-4343-2761-1 (e)
ISBN: 978-1-4343-2760-4 (sc)

Printed in the United States of America
Bloomington, Indiana

This book is printed on acid-free paper.

TABLE OF CONTENTS

ACKNOWLEDGEMENTS

First and foremost, I would like to thank the National Aeronautics and Space Administration for their support of space entrepreneurs around the world. Thanks also to Gene Meyers, Terry Martin, Mike Bain, Kimberly Campbell, Robert P. Hoyt, Chris Reed, Dr. David Livingston, Stefanie Countryman, and the rest of the brave pioneers that are continuing their efforts at the new conquest of space.

A very special thanks to Brian Thorpe for his editorial support.

INTRODUCTION:
THE SPACE RENAISSANCE

ren·ais·sance (rĕn'ĭ-säns', -zäns', rĕn'ĭ-säns', -zäns', rĭ-nā'səns)
  Renaissance
  a. A revival of intellectual achievement.
  b. The period of such a revival.

LIKE EARTHQUAKES, REVOLUTIONS CAN START in a day, but they take decades to develop. Commercial space aviation was no exception.

The 21st Century dawned with a clap of thunder. The shadows of September 11th loomed long and dark over foreign policy in the years that followed. A recession laid over the land and employment in the aerospace industry dipped to its lowest point in five decades. But out of this crucible, manned commercial space flight was born. Scaled Composites' SpaceShipOne made its inaugural flight on June 21, 2004. The commercial space age had entered into a new phase: manned flight.

Breaking every altitude record for a piloted commercial plane, the spacecraft soared into the southwestern desert sky three times, winning the $10 million Ansari X-Prize. In July of 2005, Richard Branson sighed an agreement to use the Scaled Composite design for the first commercial manned space flight company, Virgin Galactic. The technology was licensed from former Microsoft founder, Paul Allen's Mojave Aerospace Company to build SpaceShipTwo, a flight ready spaceliner. Hundreds of people signed up immediately including the venerable icon or space travel, William Shatner. A new spaceport has been built in New Mexico that will be dedicated to this and other commercial space flights in the coming years.

Despite the excitement, one cannot stop to wonder when the novelty of suborbital flight will evolve into something more significant. What would be the next logical step once tourism is established in space?

The commercial space market is evolving toward even higher goals and bigger rewards. As the future of aerospace dawns before us,

Scaled Composites' SpaceShipOne in 2004
(Image courtesy of Scaled Composite, LLC)

a commercial habitat in space is on the horizon. In this orbital facility, tourists have a destination, scientists have a test environment, and manufacturers have a pilot factory. While tourists expand their experiences, industrial scientists can create exquisite machine parts, electronic components, and wonder drugs in a weightless world unto their own. The commercial space station will provide a beachhead for governments, large businesses and entrepreneurs, and a crucible for space application development.

The benefit of a commercial space station is its zero gravity environment. It will allow liquids to combine evenly, alloys and metals to form perfectly, and circuit boards and semiconductors to reach their extreme levels of speed and reliability. Because of the quiescent environment space provides, antibiotics and antiviral drugs can be perfected for medications and vaccines. A commercial space station will also allow for the inexpensive manufacture of satellites, a storage facility for salvaging old space wreckage, a fuel depot for deep space flights, and a place for tourists observe the planet Earth and the Universe around us.

The space station can also be made into a warm and familiar environment. It could even serve as an orbiting hotel. By gently rotating the station, artificial gravity would be produced so that tourists can take a hot bath, sleep in a restful position, and dine as they would at any fine hotel in the world. The heroic days await those lucky enough to understand and define carefully the benefits of this next goal in the commercial space age.

The Vision for Space Exploration, inaugurated in January of 2004 by George W. Bush, gives the orbital space market unprecedented opportunities. Mr. Bush has mandated that America go back to the Moon, leaving the orbital market wide open for entrepreneurial efforts in space station services. The American segments of the ISS are now mostly devoted to exploration, leaving many universities and corporations without a laboratory to perform space experiments. A commercial space station can lease lab space to government-funded customers. Even NASA will find it cheaper to rent facilities in space stations than to own and operate their own. Space agencies around the world would become customers instead of landlords and superintendents, saving the taxpayers millions.

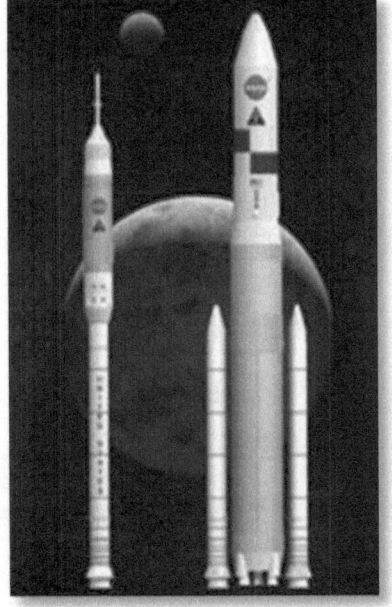

NASA's new Moon vehicles, Ares I and Ares V
(Image courtesy if NASA)

From the tragedy of the Space Shuttle, Columbia disaster this new vision was born. When the spacecraft broke up in February 2003, there were already plans for a successor space plane. Now, the focus is on creating a stack of rockets unrivaled since the Apollo project of the 1960s. These will be the foundation of an interplanetary exploration program, as well as an International Space Station (ISS) launch support system.

It also leaves big opportunities in the re-supply market. One rocket shouldn't bear all the burden of supplying the ISS. Different makes and models of rockets are needed in case of failures. Fortunately, there are plenty of companies that can fill cargo services. Low earth orbit is quickly becoming a commercially monopolized sector.

Satellites were our first commercial space stations. Their vitality is the biggest proof that space is multibillion-dollar business. Today, with GPS and radio satellites are rounding out orbital telecommunications, and now satellites have taken an even larger role in our daily lives. If a manufacturing facility could be placed in orbit, satellites could be built and deployed on the spot. They could also be salvaged, repaired, refueled, stored, and recycled. The recovery business alone is worth millions, and the savings on satellite insurance, which is 15%-20% of the launch costs, millions more.

The International Space Station (ISS) will face decommissioned in 2016-2020. What then? There are as yet no plans to build the next space station other than by entrepreneurs described in this book. The promise of microgravity is to stimulate the economies of the world and only the business sector can truly do this. As NASA flies higher, so must the businesses that support it. As we shall see in the following chapters, a commercial space station will be the financial fulcrum for the next era of lunar and interplanetary colonization.

Most space advocates can agree on the need for a commercial destination is space. It would help advance propulsion, life support systems development, and on-orbit construction methods. Once deployed, a space station would help enrich the world's economy. This book, a sequel to my first book, "The Commercial Space Age: Conquering Space Through Commerce" describes how feasible space economic development can be, and endeavors to convince those who either have the power or the means to invest, or those who are just dreaming, to see near Earth space as the future of almost every sector of the worlds' economy. Old concepts and forgotten dreams are coming back to life as the renaissance in aerospace begins.

1

COMMUNICATION SATELLITES: THE FIRST COMMERCIAL SPACE STATIONS

EXPLORER 1, AMERICA'S FIRST SATELLITE, rose into the sky aboard an army Jupiter C missile in 1958. The Eisenhower Administration could breathe a sigh of relief. The Sputnik threat had been blunted. But a new arms race had begun and no one knew how it would end. The rockets that launched the Soviet Sputnik satellite were powerful and long range. The payloads they carried could have just as easily been nuclear weapons. Satellites would soon have cameras aboard them to spy on America from space. Even if the dreaded nuclear attack never became a reality, there would be no real peace in the world for decades to come.

Some, however, watched Explorer 1 without thinking about military applications at all. Some saw great scientific observatories in the sky. Satellites could measure the upper atmosphere and magnetic field of Earth. And by looking back down at Earth's surface from space, weather patterns could be predicted with unprecedented accuracy, and the mysteries of our atmosphere and ocean currents could be solved.

Many dreamers saw beyond the atmosphere. Observations of the sun could be achieved with great precision from a satellite. The inner planets could be mapped and surveyed, and a great astronomical

Early concept of an applied technology satellite.
(Image courtesy of NASA)

observatory could be orbited above the distorting effects of Earth's atmosphere to study the stars. Someday satellites could orbit the Moon to see the far side, or even maintain orbits around planets and moons of the outer solar system.

Others believed that space exploration would usher in a great age of computers and robotics. Because of the inherent perils of space, robots could extend the human senses, calculate at amazing speeds, and could sacrifice themselves for the sake of discovery. The development of powerful, compact computers, tailor made for space flight, would be needed to accelerate for the Space Age to its full potential. Soon a new generation of computers would be a part of everyday life. They would be all around us, serving our every convenience.

But most of us saw the beginnings of manned space flight taking shape. By the 21st century, there would surely be space stations and colonies on the Moon. Business people would travel through space as easily as they do through the air. Every major city on Earth would have a spaceport, and the Moon would the next industrial and recreational Mecca.

But what actually happened was very different. Instead, space became a broadcasting haven for telecommunication companies. The globe was soon crowned with unmanned space stations, sending signals to subscribers who were listening, viewing, and talking with their clients. In line with forecasts, this space race ended in commercial success for all nations, and private industry reaped the spoils, with revenues of up to $100 billion per year. The satellite industry did not develop organically, but through the balanced cooperation between government and industry.

In the early 1960s, the pace of research in satellite development was brisk. After Sputnik's debut, the United States began flying inflatable Mylar relay satellites. Broadcasting companies, well aware of the potential of orbiting television antennas, began doing research of their own. In 1962, the first near Earth orbiting satellite was deployed (Telestar 1) which made telephone transmissions to Europe and the US.

A nexus in technology had emerged that made satellite commerce really possible. The first was the Traveling Wave Tube, a powerful signal amplifier to allow the weak satellite transmission signals to be heard on the ground. The second was Dual Spin Stabilization, a way to keep satellites from tumbling and wobbling in orbit. These two developments, together with new and powerful receiving stations on the ground, made satellites practical for the commercial broadcasting business almost overnight.

But in 1962, when Telstar 1 was launched, all the big decisions still had to be made. Even in those murky days, President John Kennedy quickly recognized the social and economic significance of the new technology and embraced it. He released the following statement in the summer of 1962 after Telstar was launched:

> "The successful firing and subsequent operation of the Telstar communications satellite is an outstanding example of the way in which government and business can cooperate in a most important field of human endeavor. The achievement of the communications satellite while only a prelude already throws open to us the vision of an era of international communications. There is no more important field at the present time than communications and we must grasp the advantages presented to us by the communications satellite to use this medium wisely

and effectively to insure greater understanding among the peoples of the world."

Great opportunities for peaceful use of satellites were becoming tenable, but it would be very expensive. How was it to emerge from the shadows of the defense department and corporate monopolies? How was it to evolve into a peaceful, affordable innovation for all to use?

The American Sputniks

AT&T Bell Labs soon had a grand idea to extend its monopoly of international telecommunications. It would deploy fifty satellites that would orbit the Earth in a continuous "constellation" of radio relays. The company would build the satellites, and *even pay NASA* to launch them. They would track them from their Andover, Maine headquarters and proceed to send signals to their customers in Europe.

The first launched was Telstar 1, a wobbling, beeping, geek of a satellite deployed a few thousand miles up in a medium orbit. Television and

Jupiter C missile carrying the first American satellite, Explorer 1 into space. *(Courtesy of NASA)*

telephone services were the markets. As early as 1962, only five years after Sputnik, a private firm was building, launching, and operating a new platform for global communication.

Soon Europe heard the news of the "American Sputnik." A space age had come to the west at last, but the US government was as wary of AT&T as they were of the Soviets.

Not to be outdone by AT&T, Hughes Aircraft flew a satellite that would demonstrate that commercial broadcasts could be made over a large area using only one spacecraft instead of fifty. If a satellite were deployed at a very high altitude along the equator, it would hover over the same hemisphere in a stationary position. In mid-1963, Hughes Aircraft and NASA launched the Syncom 2 (Syncom 1 did not make it into orbit) that deployed itself at 23,500 mile altitude, and hung motionless in a geosynchonous orbit over the western hemisphere. It was spin stabilized to cancel out wobbling, and kept the transmission beam aligned. This was the prototype for the broadcasting satellites we use today.

Shortly after the Telstar's launch, the Kennedy administration made the announcement that legislation would be drafted and passed by Congress to aid in the deployment of satellite technology around the world. The Communications Satellite Act of 1962 was the most important piece of domestic legislation since the Federal-Aid Highway Act of 1956. In this document, was the legal groundwork for the $100 billion dollar per year industry we have today. It mandated that a global satellite network be put into place as soon as possible and that all interested parties would have fair access to the system.

To head off an AT&T monopoly, the US Congress created a single company, the Communications Satellite Corporation (COMSAT) in 1963, to regulate and market all participants in satellite communications for the US. The company would represent the US participants in what was soon to become a global association of nations who would benefit. No monopoly could quash the efforts of smaller carriers. The playing field would be level.

In 1964, a consortium of over 100 governments formed the global commercial telecommunications carrier, Intelsat Corporation. It was to provide services to all areas of the world and contribute to world unity. Each nation had its acting marketing member (America's was Comsat) who would participate in the planning, construction, ownership, management and operation of the network. Comsat (60% holder of the business in those days) would be supervised by the Congress to assure fair competition in the market. Now the satellite communications business could be regulated under the government's thumb at least until the market matured.

That happened faster than expected.

Here were two models of how industry drove NASA to create spacecraft for apolitical, financial reasons. In AT&T's case, NASA was *procured* for their services and equipment. In the Hughes case, NASA *assisted* in the development of the spacecraft. The satellite broadcasting market was very much in place at that time. In fact, before these projects had begun, France, the United Kingdom, Brazil and Japan had already created ground stations to receive the signals from the coming fleet of spacecraft. A satellite could last twelve years in a high orbit and could service half the world with television signals and telephone calls. If the financial payoff was quick enough, business concerns could leapfrog over government bureaucracy.

Intelsat has continued to be the most ubiquitous satellite company in the world. The fleet has over 50 satellites in geostationary orbits, strung like pearls around the belt of the Earth; with the entire civilized world is its customer. This all grew out of the 1962 Congress who grasped the economic potential commercial space promises us.

By 1973, NASA stopped all research in telecommunication satellites, leaving it to private industry that had the financial incentive to develop the technology to greater heights (NASA later resumed research in the early 1990s). For 25 years, all was stable in the satellite world. But the Intelsat consortium was in for a surprise.

Mobile Satellite Service Emerges

World commerce was on the move. Soon mobile satellites were deployed to provide instant communications on the rollicking decks of military vessels and airplanes. Maritime communications required the satellites be very powerful, because mobile receiving antennas needed to be small and portable, not like the giant dishes typical in those days. The then Soviet Union, a member of Intelsat, suggested that a satellite be built that was completely devoted to air and sea traffic communication. This would allow for partial Soviet sponsorship of the project. In 1976, the international Marisat satellite was launched. Even in these Cold War depths, the international commercial space model survived because it was so universal and practical.

A new company, the International Maritime Satellite Organization or Inmarsat, was founded in 1979 and began specializing in maritime and aerospace industries. It owned three satellites, an initial customer base of 1000 ships and an operating communications network for ocean going vessels, trucks, or planes, well away from wires and cables, could use relay satellites. Using the same business model as Intelsat, it started off as consortium of financially contributing international members, and later became the mobile satellite equivalent of Intelsat.

The Inmarsat Company now has an 11-satellite constellation that is tapping into middle-class consumer markets for mobile Internet service. The company has recently tied together many developing nations in a broadband, desktop terminal based network, called the Broadband Global Area Network (BGAN). The service will be accessible with small, lightweight satellite terminals that can be connected quickly and easily to laptop computers.

The New Satellite Wave

Back in the early eighties, CNN and ESPN became serious contenders for network TV market share. Cable television needed satellites to fill in gaps and provided service where cables could not reach. You could run cables just so far into the wilds of undeveloped countries before services attenuated. Providing service for the "last mile" became an important selling point for satellites providers in the years to come. Since cable companies needed a way to broadcast to areas of the world that had little in the way of infrastructure, a market emerged for a private firm to take the initiative. The result was space entrepreneurialism at its best and bravest.

In 1984, Rene Anselmo, President of the Spanish International Network (SIN), a company dedicated to the development of Spanish-speaking television stations, founded the satellite company, Panamsat. Using his own money, Anslemo bought a cut-rate satellite from RCA for half price ($45 million) and launched it with Europe's then temperamental Ariane booster rocket for a low price. In 1988, the launch succeeded and PanAmSat was up and functional with CNN and ESPN coverage worldwide.

PanAmSat was the first private company to compete with the mighty Intelsat. The company seized the opportunity to create a new market. The new satellites would complete the final links in programming that would dominate television to this day. The Galaxy satellites, as they were named, would provide the constant coverage of world news and sporting events reported as they happened wherever they happened. PanAmSat was created at just the right time to take advantage of the next generation of broadcasting.

Timing was everything. A few years after the launch of the first PanAmSat satellite, the Soviet empire fell, and shortly thereafter, the Gulf War began. There were big broadcasting revenues to be made. PanAmSat expanded operations by buying more satellites through the issue of junk bonds. Though a few satellites failed before operation, PanAmSat forged ahead, eventually merging with Hughes Communications in 1996. Focusing mainly on television markets, PanAmSat has now broken into the HDTV markets where the next trend in broadcasting is headed.

By 2005, the company owned 24 satellites, with assets of 6.3 billion. Focusing on video, PanAmSat emerged when the technology and the markets were in need of a gap-filler in worldwide coverage. Where would CNN or ESPN be without satellites to extend to their "last mile" service? Out of business, probably.

This is an example of a space business model that was completely self-financing. It did its job of filling in the gaps in the market just as cable TV began to explode. In 2005, Panamsat and Intelsat merged into the largest satellite company in the world. The next largest, SES Global, has one-third to one-half the capacity. Intelsat's strengths, which are in telephony and data, will be combined with Panamsat dominance in video centric services.

Satellites Enter the Living Room

As the private sector space economy began to expand in the 1980s, so did satellite communications technology. But it wasn't all milk and honey for the business world. In 1976, a former NASA planetary scientist and Stanford University Professor, H. Taylor Howard, devised away to tap signals normally reserved for wealthy subscribers. He built a large twelve-foot, spider web of an antenna that could be placed on a garage roof or in a back yard. These relics can still be seen today. By 1980, 5000 operating dishes combed the skies. No de-scrambling technology was in place, as it is with cable and satellite TV now. Anyone with enough skill could build a dish and get free satellite broadcasts for hundreds of channels. You could even steer the dishes to link up with more than one satellite. The price tag was about $10,000.

Cable companies, which broadcast via satellites to various regions, realized that a substantial number of users would soon be getting TV broadcasts for free. They lobbied Congress to allow them to scramble signals to prevent non-paying users from intercepting programs. The Cable Communications Act of 1984 was passed through Congress that provided the legal framework for cable companies to encrypt their transmissions and prevent satellite-based non-subscribers from accessing cable stations. This "good" regulation was an effort to better diversify cable TV programming and help the sector grow without piracy.

Direct broadcasting satellites ushered in affordable commercial space services to mainstream consumers without piracy.
(Image courtesy of NASA)

Though this was an unpopular development, what else were cable broadcasters going to do? Despite the efforts to foil piracy, illegal de-scrambling continued rampantly. Things looked like they were unraveling when a good portion of the general public relished the opportunity to pirate overpriced services. The answer was actually to make satellite TV more accessible to the consumer at a "cable company" price. At the same time, a powerful new generation of satellites could actually beam signals that even tiny household dishes on garage roofs could receive.

A new company emerged, BskyB, led by Rupert Murdock in the United Kingdom. This business offered TV service through small gray dish receivers that most middle-class consumers could afford, and even install themselves. They were simple and fixed, and came with a tuner box to decode the signals from space.

The industry took off like never before. Echostar launched a competing satellite network (Dish Network) in 1996, and Hughes Satellite bought out Primestar and USSB by 1999. In the same year, President Clinton signed the Satellite Home Viewer Improvement Act (SHVIA) into law, which allowed consumers to receive their local broadcast signals via their direct broadcast satellite system. This truly made satellite TV competitive, and was the deciding factor for many subscribers to switch from cable to direct-to-home TV.

The age of piracy faded away, and the big black dishes were replaced with little gray ones. Two American companies, both rooted in big government satellite programs, were fiercely competing with each other and with cable. Dish Network and Direct TV grew geometrically into becoming the two giants of the industry by the year 2000.

A dozen satellites were launched dedicated to direct-to- home TV. From Canada to Japan and the Arab nations, satellite subscribers reached into the millions. There is no inhabited place on Earth where satellite TV is not somehow reachable. At the time of this writing, over 20 million Americans subscribe. Additionally, many underdeveloped counties, such as Iraq and Afghanistan, suddenly have the "instant infrastructure" to broadcast news and other programs.

These positive steps that Congress signed into law, helped grow the satellite industry without undue delays. It shows that proper regulation has worked very productively in space commerce and has every precedent to do so in the future. With aggressive markets driving away government hesitancy, the industry grew to its present level in one generation.

The Consistency of Low Earth Orbiting Satellites

During the 1980s, the high cost of space access was just as crippling as it is today. Satellites cost $2000-$4000 per pound to reach orbit. But shortly before the Challenger Disaster in 1986, Orbital Sciences Corporation appeared on the radar screen. They were a recent entry in the aerospace business community and had planned to use satellites in a way that was reminiscent to the AT&T Telstar 1 constellation. To make this affordable, they also needed to deliver as many satellites as they could with the fewest launches possible. To accomplish this, Orbital Sciences had to create a delivery system that was revolutionary.

The first customer of Orbital Sciences was Canadian mobile telecom company, Teleglobe. They were to buy the transponder rights to low earth orbiting satellites once they were deployed. A new company was formed called Orbcomm, and it was a pioneer in low cost commercial space projects.

This constellation of 30 satellites, circling in non-geostationary orbits, 500 miles up, sweep over every continent on Earth and provides two-way communication to mobile and stationary subscribers. The customers were those who needed to track the position and operation of vehicles and machines and monitor and allocate their productivity. Ships and trucks could all have their operational status metered and their exact positions tracked by satellite. This allows operators to coordinate their assets more efficiently.

The Orbcomm model is an example of creative financing and engineering of a space commercial product, complete with a sound launch and marketing strategy. The only problem was that it was a little ahead of its time. The development period was grueling, waylaid by technical and start-up problems. But Orbcomm did eventually fly and provide the services it had promised.

Two other companies then joined in on the LEO market with telephone services: Iridium and Globalstar. In both cases, the strategy was to make them low orbiting, affordable, and accessible from anywhere on Earth. Customers didn't need "a clear view of the southern sky" to use them. You could be in the depths of a rainforest, on an arctic plain, and still get service.

Though all LEOs worked, they did not, however, live up to their promises in the financial department. All have been in and out of bankruptcy due to weak subscriber numbers, but are still around having survived their refinancing deals. Now completely recovered, all three LEOs are planning on new generations of spacecraft that could better compete with Inmarsat broadband services.

But there is one more LEO success story. Sirius Radio uses a low Earth-orbiting constellation that is quite a popular item now. With no commercials, and a steady signal, the change from the tired radio formats to a refreshing "cable quality" format is unique. Sirius uses three low earth orbiting satellites drifting in paths that are 15,200-29,280 miles above the earth's surface. Since they do not hover in the sky like geostationary satellites, they take turns broadcasting in elliptical eccentric orbits above North America and Canada. There is at least one over the U.S. at all times. (The XM radio satellites are in geostationary orbits.) To keep in constant contact with listeners, they transmit to ground receivers first and relay the signals to repeater stations throughout the U.S. Though Sirius still has a way to go before paying off its debt, it continues to make its quarterly subscriber goals.

Low Earth Orbiting (LEO) constellations were once considered to be a failure, but this has now changed. Hybrid LEO and GEO connectivity seems inevitable. Bottlenecks can be relieved and patches created by the redundancy provided by LEOs in orbit over large cities. Should one LEO satellite fail, a financial catastrophe can be avoided because a spare, already in orbit can be activated.

In December 2006, Globalstar and Alcatel announced a new medium Earth-orbiting constellation that will allow for voice, data, and broadband access anywhere in the world. The plan is to install 48 new satellites that will have a lifespan of 15 years. It will provide service for 120 countries over six continents. Since Globalstar is a mobile service, they can provide broadband communications anywhere, anytime for anyone.

Omnipresent Eyes in the Sky

Imaging satellites (also called remote sensing satellites) take pictures of the earth from space. After the September 11th attacks, they have had resurgence. America's pre-emptive strategy required more and more intelligence and faster ways to process data. Imaging spy satellites can and always have saved the American taxpayers billions be verifying whether or not their enemies have weapons and facilities that threaten U.S. interests. By knowing what an enemy has or does not have in their arsenal has allowed for more adroit foreign policy.

The rationale for spy satellites dates back to the Lyndon Johnson Administration in the 1960s. For many years Americans feared that the Soviets were making nuclear missiles as fast as sausages. But images from space showed that the suspected Soviet arms build up did not match up with what the spy satellite were seeing. As it turned out, the satellites showed that there was no real gap in the number of ICBMs the US and the Soviet Union had deployed. Based on the images taken from space, the National Security committees did not have to commission the building of more and more missiles or bombers. This kind of surveillance saved the taxpayers billions in false intelligence leads.

As Lyndon B. Johnson said a gathering in 1967,

"I wouldn't want to quoted on this. [But] we've spent $35 or $40 billion on the space program. And if nothing else had come out of it except the knowledge that we gained from space photography, it would be worth ten times what the whole program has cost. Because tonight we know how many missiles the enemy has and, it turned out, our guesses were way off. We were doing things we didn't need to do. We were building things we didn't need to build. We were harboring fears we didn't need to harbor."

If the satellites do show enemy hostilities looming, targeting and damage assessments can be made much easier from the safe haven of space instead of from highflying aircraft. During Operation

Freedom in Iraq and Afghanistan, it was essential to know exact target coordinates and to guide missiles only to the targets in that controversial war. The effectiveness of bombings can be assessed by satellite images so that no one has to send the missile again just to make sure the target was hit, incurring civilian deaths in the process. Satellite imaging methods allow for surgical targeting, attacks and validation, and have contributed to effective in real-time tactical planning and situational awareness, and can only be expected to play a growing role in the future.

One cannot underestimate, or even properly measure, the reduction of casualty rates due to satellite assistance on the battlefield. Imaging satellites are as significant for defense as nuclear bombs are for the offense. Battlefield operations revolve around them. Contrary to their military counterparts, commercial imaging companies have had a rough time. They have to compete with airplane photography under peaceful civilian skies.

The LANDSAT imaging satellite was first launched in 1972 as the Earth Resources Technology Satellite and proved itself right away in resolution and scope. A total of seven have been flown as of 2005. They were placed in a polar orbit 570 miles up and can give high-resolution images of any of Earth's surface features. Some of the applications for these images are:

Agriculture, Forestry and Range Resources

- Discriminating vegetative, crop and timber types.
- Monitoring and measuring crop and timber harvests.
- Precision farming land management.
- Determining soil conditions & associations.
- Determining range biomass & health.
- Monitoring irrigation practices.
- Assessing wildlife habitat.
- Monitoring and mapping insect infestation

Land Use and Mapping

- Classifying land uses and land capability
- Cartographic mapping and map up-dating

- Monitoring urban growth
- Regional planning
- Mapping transportation networks
- Siting power transmission networks
- Planning waste disposal sites, power plants and other industries
- Mapping and managing flood plains
- Tracking socio-economic impact on land use

Agriculture, Forestry and Range Resources

- Discriminating vegetative, crop and timber types.
- Monitoring and measuring crop and timber harvests.
- Precision farming land management.
- Determining soil conditions & associations.
- Determining range biomass & health.
- Monitoring irrigation practices.
- Assessing wildlife habitat.
- Monitoring and mapping insect infestation

Geology

- Mapping major geological features
- Revising geologic maps
- Classifying rock types
- Mapping volcanic surface deposits
- Identifying indicators of mineral and petroleum resources
- Producing geomorphic maps
- Mapping impact craters

Environmental Monitoring

- Monitoring deforestation
- Mapping and monitoring water pollution
- Determining effects of natural disasters
- Tracking oil spills
- Mapping and monitoring lake eutrophication
- Monitoring volcanic ash plumes

- Assessing and monitoring grass and forest fires
- Assessing drought impact

LANDSAT was owned and operated by the U.S. government for the study of Earth's physical and biological structure, but the commercial applications are obvious. The Carter administration began to try to commercialize it in 1979 with the understanding that market may take a decade to develop. The Regan Administration hurried the process, and perhaps birthed the baby too early.

Using an "if you build it, they will come" approach, LANDSAT became the antithesis of the Marisat or Intelsat programs, and an ominous lesson to commercial space companies everywhere.

In 1985, an experimental cooperation attempt began between NASA and Earth Observation satellite Company (EOSAT) to operate the LANDSAT system under a ten-year contract. If successful, anyone would be able to buy images of the Earth. The problem was that they were so expensive, and only the government could afford them ($1000-$4400 an image, depending on the processing involved). The government was now the only customer and was overpaying for its own pictures. In 1994, Rep. George Brown, Chair of the House Committee on Science and Technology wrote Vice President Gore describing the Landsat Program as a financial "shambles".

But due to civil responsibilities in an era of terror and environmental controversy, the U.S. government bought out the company by the end of 1995 and now runs the operation itself. This can show what can happen to a failed Government/commercial partnership that does not have a fully developed market to sustain itself. Now LANDSAT 7 orbits the Earth as a government funded operation. This failure occurred because the market was premature and there were not enough commercial customers that could afford the service. The market changed in just a few years.

By the 2007, two imaging satellite companies emerged, DigitalGlobe and GeoEye. All use satellites in a low polar orbit about 300-500 miles up. (GeoEye uses the IKONOS, IRS, and Orbview-2 satellites, and DigitalGlobe uses the Quickbird). These three companies have standard markets in civil engineering, national security, agriculture, and resource exploration, but also have begun to help the media press cover disaster areas. By using imaging and telephones services, the low earth orbit market benefited enormously in the wake of hurricanes Katrina and Rita that devastated the American gulf coast. By facilitating humanitarian aid communications in a complete infrastructure breakdown, low Earth orbit satellite services were the some of the only alternatives rescue planners had available.

Space borne imaging is always competing with aircraft imaging. In wartime, satellites become necessary as they can stay out of enemy missile range. In peacetime, satellite images can provide detailed panoramic views aircraft cannot. Low-resolution panoramic views of the Earth and atmosphere require an orbital satellite's high altitude vantage point. The media market for images will no doubt grow because of the time issues involved in requisitioning aircraft pictures. News stories now often include high-resolution satellite shots of natural disasters, crime scenes, and even archeological discoveries that require fast internet coverage.

GPS Auto Guidance Comes of Age

The Global Positioning System, a fleet of 24 satellites in medium orbit 10,000 miles up, is now a staple of everyday life. Automobile companies can now incorporate GPS into cars and trucks, along with satellite radio, broadband Internet and TV. The uses for GPS are too numerous to mention here, but it is by far the most successful use of a satellite constellation ever devised. Europe soon will place its own $3.6 billion Galileo version into medium Earth orbit for civil uses in rescue, airline traffic control, law enforcement, etc. and for the general benefit of society. Russia and China now have there own constellations.

The future of GPS is not in locating the positions of vehicles, but actually guiding them. One such GPS steering system is already on the market. This NASA spin-off is a device used to guide farm vehicles with a GPS-fed "driver". The RTK Autosteer™ was released in 1999 to farmers around the U.S. who need to drive tractors in neatly defined rows for optimal seeding, watering and fertilizing. The Autosteer uses GPS to automatically correct for errors, thereby enhancing the accuracy of the task. This precision increases the yields of crops. The Autosteer also adjusts for vehicle performance changes, varied terrain conditions, and unexpected encounters with people and objects.

Automated farming is a very tenable goal in GPS guided equipment. At the University of Illinois, engineers now are developing robotic farmer vehicles that actually perform plowing, fertilizing, dusting, and harvesting crops using GPS and laser (range finder) guidance systems. Since the cultivation of ethanol producing crops will become more expensive in the future, auto-guided farming becomes attractive to the labor-intensive alternative.

Satellite Commerce: Present and Future

What will communication satellites be like in the future? As far as one can see, it will all be evolutionary: bigger, more powerful platforms. Basically, giant satellites that provide broadband TV, broadband internet, radio broadcasting, precise GPS services, and higher resolution imaging, is where the industry is headed. Mobile services are becoming more and more important, so more ubiquitous receiving stations complement the orbital services.

The first requirement for a big satellite is generating electrical power. This is based on the size and quality of the solar panels. Since batteries or nuclear power sources are not considerable in space, power will continue to be solar, and more and more efficient solar panels are being developed. Solar panels will soon be reaching the 40-50% efficiency range with up to 18kw per satellite generated.

The second is transmitting power. The more intense or concentrated the beam sent to Earth, the higher the transmission of data. Most satellites have the ability to spot beam a signal to a specific market center where a certain event needs to be viewed. Spot beaming uses a transmitting dish to narrow the focus of the transmitting beam and increase its intensity. Small direct-to-home dishes rely on powerful beams emanating from large satellites.

Satellite power is climbing rapidly. In 1962, the Telstar 1 had 1 watt of power and sent faint signals that only large powerful receiving dishes could pick up. Today's satellites are all about high bandwidth, high definition and speed. XM radio satellite (3 kilowatts) was one of the most powerful ever launched at the time 2001, but a new line of European satellites will eclipse 20th century standards in electrical generation. Downloaded data becomes more concentrated and faster, whether it is for broadcasting, faxes, Internet pages or video. The New SpaceWay (Direct TV) satellites, originally designed for broadband Internet, have been repositioned for high definition television. In addition, new satellites will have as many as 190 transponder connections that can be leased out at all once to a variety of customers. Normally satellites carry only a fraction of that capability.

Bigger has become better. The European Space Agency together with satellite manufacturers, Centre National d'Etudes Spatiales, EADS Astrium and Alcatel Alenia Space are building the Alphabus chassis. This is a for huge, 6000-8000 kilogram satellites generating 12-18 kilowatts of power. These can handle all business and commercial applications that require broadband, as well as HDTV, telephones, etc. A virtual "office in sky" for anyone who can lease the transponders.

Loral Space and Communication Systems has already built the heaviest satellites ever flown into high Earth Orbit (6505 kilograms). The IPSTAR-1 was launched by the Arian 5 rocket in the summer of 2005. This spacecraft generates 14 kilowatts of power. The transmission rates are 8 megabytes per second, 2 1/2 times faster that of the average household cable line. The lifetime is expected to be 12 years.

Mobile consumer applications are expected to grow vigorously. Not just for ships and trucks, but for casual travelers. Campers and mobile homes equipped with broadband Internet are within the reach of middle class budgets and are now available. Though thousands of dollars are needed to get going, if you're a serious mobile home nomad, you're living standard will go way up with access to satellite radio, broadband internet, fax machines, HDTV, theft protection, telephones and GPS navigation to make your mobile home a paradise of convenience. Consumers can even run a home business out of their recreational vehicles.

Swinging on a Star: Reducing the Cost of Satellites

Space Tethers Incorporated is developing a technology that could lower the cost of orbital maneuvers and the price of satellites overall.

The company manufactures cables made from woven strands of tough polymer fibers. They can connect satellites in tight formations or lift them from one orbit to another. These could reduce the need for the impulse engines used to move satellites around once they are in orbit, lowering the cost of building them.

One of the expenses in satellite deployments is using a kick motor to transfer satellites form on orbit to another. Satellites are launched with up to four rocket stages, each bringing the spacecraft to a different altitude. These motors can malfunction and can strand spacecraft in useless orbits. Sometimes a Shuttle astronaut can retrieve them, others are so high that nothing can save them. Insurance rates are very high for this reason, but by using tethers to lift satellites into higher orbits, a lot of hardware can be spared. Tethers can be used to keep satellites in tight formations. By connecting them in a spider web of cables, constellations won't drift apart over time.

Electrically conductive tethers can capture energy from the Earth's magnetic field. By using this trapped energy, the cables act as electric elevators for satellites to move up and down. In fact, the ISS can use this technology to maintain its orbit instead of using the expensive chemical rockets as it does now. Conceivably, a large solar power station could be boosted into higher orbits by these magnetic lifts. Tethers become electrified naturally and would need no solar power to operate.

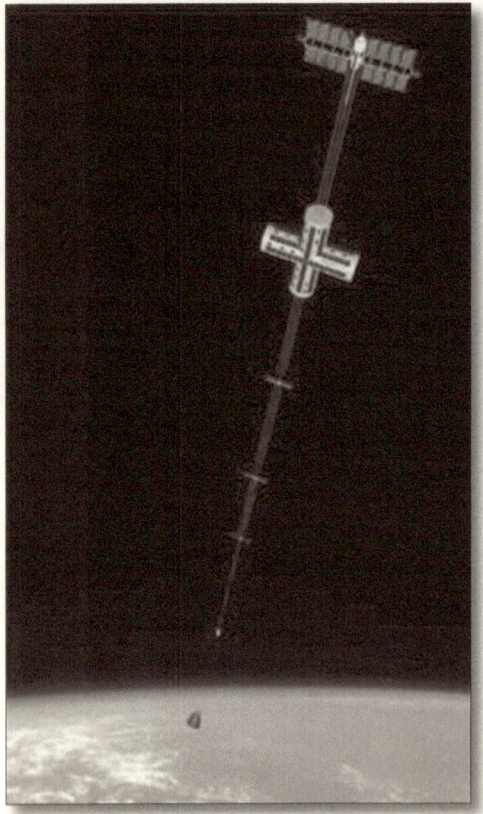

A satellite being lifted by an electrically charged tether to a higher orbit. *(Image courtesy of Tethers Unlimited Incorporated)*

Not only could satellites be tugged to various orbits without using expensive and dangerous rockets, but they could also be de-orbited when they expire so that they do not clutter up space.

How Satellites Can Use Tethers

1.	**Momentum Transfer** - tossing satellites from one orbit to another. This includes de-orbiting satellites that would otherwise pose a danger to other spacecraft
2.	**Powered Transfers** - conductive cables can use Earth's magnetic field to hoist satellites up to higher orbits.
3.	**Formation Flying** - tethers can connect satellite groups together in tight formation to keep them from drifting.
4.	**Tugs** - spacecraft (chemically powered) can pull other un-powered spacecraft around without the need for docking. Tethers could be used for a Moon tug.
5.	**De-orbiting** - tethers can by used to grapple space debris for salvage or disposal.

Business Models for Space Entrepreneurs that Actually Worked

What emerges from this historical glimpse are three business models that have proven effective in financing the high price of space commerce. The "if you build it, they will come" method is *not* one of them.

1. Model #1:
Government-Industry Cooperation - Certain projects that NASA cannot or will not pay for are phased into a commercial effort. NASA and other government agencies may need to remain as customers for the effort to survive at first. Affordable pricing and skillful marking is key for the customers to respond. They must understand why they need space services.

2. Model #2:
Privately Financed - A private entrepreneurial company finances whatever money is needed with the knowledge that the time is right. Customers are waiting before the service is ready.

3. Model #3:
Spin Off - Licensing current or surplus space assets, designs, patents, and hardware.

Business Model	Companies that used them	Advantages	Drawbacks
1. Government-Industry Cooperative	Intelsat/Comsat, United Space Alliance, Inmarsat, Eosat	Lower initial investment, government as customer.	Government funding is politically dependent. Lower profit potential. Very high initial prices.
2. Privately Financed	PanAmSat, Orbcomm, GlobalStar, Iridium,Virgin, Galactic, Orbital Recovery, XM, Sirius, Bigelow	High profit potential.	High initial investment, high risk due to market and technology shifts, highly competitive with ground-based services some of which are government financed.
3. Spin Off	GPS equipment providers, solar panel producers, medical device manufacturers, etc.	Lower initial investment.	Very competitive. Proportionately lower profits

With satellites now connecting the world with music, data and cell phone networks, the foresight and rationale of Congressional regulators back in the 1960s couldn't have been better guided. Rather than hampering the development of satellites, the U.S. government greatly aided it. However, unlike the satellite business, manned space technology did not evolve rapidly at all. When human beings are launched into space they need to breath, drink, eat and work. Life support systems are heavy, expensive and harder to mature if there is no wide market for them. This is why space utilization is so expensive, and why only unmanned services have flourished.

But investors can take comfort that no technology can replace the human experience of traveling around the globe. In the near future, a private company will place a habitable space facility in orbit for tourists and scientists to enjoy the experience of really being in space. The pent-up demand for space travel among baby-boomers will burst. In the following chapters, we will see how affordable new commercially funded habitats will be available for scientists who cannot book experiments on the International Space Station. There are many pieces already in place for this new era to begin:

1. A government space program anxious to devote resources to the Moon and Mars.
2. Shuttle component manufacturers who need to continue with legacy hardware and tooling to keep their companies profitable.
3. A need for the commercial aerospace industry to keep its workforce growing and competitive with the rest of world.
4. An eager public that is impatient with NASA's politically restricted pace.
5. A host of science experiments that cannot be performed on the ISS that need to be revived on a commercial space station.
6. A willingness for NASA to license its products and services.
7. The on-orbit development of national security and environmental protection applications for the public good, such as surveillance and solar energy.
8. Free GPS guidance technology available for on-orbit guidance applications.

9. Rapidly evolving space transportation systems now competing for the orbital supply markets.
10. Orbital and sub-orbital tourism remains alive and well despite the high price tag.

The Russian and American modules on the International Space Station are still performing commercial science and housing tourists. The Europeans and the Japanese will start up their modules in the next few years. As the renaissance in transportation and energy grows into new and fantastic forms, space age investment opportunities will become clearer and safer. The next generation of space stations will be as accessible as never before.

2

SALYUT TO ISS:
THE HISTORY OF MANNED
COMMERCIAL APPLICATIONS IN SPACE

AS THE NEW MILLENNIUM DAWNS, commercial research in space still remains on the ragged edge of applied science. Results are stretched painfully thin and progress is often cut short just as breakthroughs appear on the horizon. But killer applications still lurk hidden against the stars. Emerging from the biological world or from the energy field, the first practical applications from microgravity research will eventually come. The mission of commercial space stations will be to make human laboratories and living spaces affordable, and to pick up where the International Space Station leaves off. It will also provide an orbital commercial hub that will enable trade and accelerate clean transportation and energy technologies.

What would a commercial space station look like? It might be round or elliptical, wheel shaped or square, with solar panels extending from every part of it. The station would be in a low orbit, 400 miles up, and might rotate to provide artificial gravity for the people living in it. It would be visited by automated and manned vessels bringing in supplies and new customers, with rescue pods at the ready in case of emergencies. If the space station spins to provide the full force of Earth's gravity for the passengers, it would need to be large, 500- 1500 feet in diameter, and rotate two to three times per minute.

Present and Future Concepts of Privately Financed Space Stations

 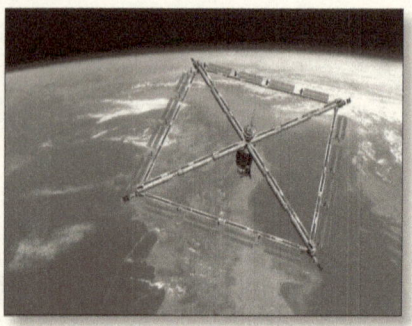

Genesis 1. Prototype non-rotating inflatable for on-orbit research is currently operating (unmanned) in orbit. *(Image courtesy of Bigelow Aerospace)*

Mature designs include square and round models that rotate to examine effects on humans and plant life. *(Image courtesy of Space Island Group)*

What kind of activities would the customers do there? The station would allow scientists, researchers, tourists, filmmakers, educators, diplomats, and business leaders to pursue their careers at a new level. From bench chemistry to show business, activities would vary only with the limits of creativity. Whether the customers are actually on-orbit or controlling events remotely from Earth, they can experience advanced medical breakthroughs, electronic wonders, and exotic sports. The artistic world will follow with new kinds of film and theater, dance, exotic sculptures only possible in microgravity studios.

Potential Commercial Space Station Clients

Client Category	Activities
Large chemical, electronic, biotech companies	Product testing, development and light manufacturing of metals, pharmaceuticals, ceramics, microelectronics.
Civil Government customers	Product research and development for space flight proficiency, including medicine.
Universities (by grant)	Pure science
Wealthy individuals	Tourism
Television and film studios	Music video, education, reality shows, movies

A nexus between lower priced space transportation, high tech materials, and a burgeoning space tourist industry is evident today. Scaled Composites', SpaceShipOne is an example of the human-rated transportation systems that are emerging. The Bigelow Aerospace's inflatables (discussed later) are examples of how cheap, sturdy and affordable new spacecraft construction materials can be, and the succession of five tourists who visited the International Space Station over the past six years are snapshots of the market demand for commercial astronautics. These developments are converging in what seems to be the crossroad marketplace: a commercial space station.

Since the years of the Eisenhower administration, every U.S. President has created his own space policy around whatever priorities the country has had at the time. George W. Bush was no exception, but no one expected such a dramatic change to occur. His is aggressive and comprehensive, and has garnered much bi-partisan support. On January 4, 2004, he announced the new Vision of Space Exploration. The main tenets are:

1. Finish the International Space Station by 2010, honoring international partners fully until 2016.

2. Retire the Space Shuttle by 2010 and replace it with two types of expendable rockets and a crew exploration vehicle (CEV).

3. Return people to the Moon by 2015 and stay there, Mars sometime in the future.

4. Welcome international cooperation.

5. Invite the business world to supplement space services.

The interesting part of this new vision is not what it intends to do, but what it avoids doing. The New Vision does not concern itself with space stations and commercial microgravity research, but in the exploration of the Moon and Mars. What it implies --and I believe intentionally -- is that near Earth space is now available for private companies to utilize. By concentrating on lunar and Mars missions, the New Vision opens up low Earth orbit for a renaissance period where all the old dreams, conceived of at the dawn of the space age, can become real.

American President	Priorities of the Day	Supported Manned Missions
Dwight Eisenhower	Defense. Catch-up to Soviet Sputnik.	
John Kennedy	Moon Race. Peaceful use of space, Soviet cooperation.	Mercury, Gemini, Apollo
Lyndon Johnson	Moon Race.	Apollo, Skylab
Richard Nixon/ Gerald Ford	Détente. Reduction of costs.	Space Shuttle, Skylab, Apollo- Soyuz
James Carter	Defense, cost reduction, détente with Soviets.	Space Shuttle, Apollo- Soyuz.

Ronald Reagan	Defense, commercialism, revitalization.	U.S. manned space station, Freedom, commercial applications.
George Bush, Sr.	Revitalization, consolidation, exploration, cooperation in space with Russia.	U.S. manned space station. Space Exploration Initiative
William Clinton	Cost reduction, cooperation with Russia.	International manned space station, next generation Shuttle
George Bush, Jr.	Exploration, commercialism, defense.	Vision for Space Exploration. Finish ISS, Space Shuttle decommission/ replacement. Return to Moon and then Mars.

NASA will fund the ISS until 2016. By that time, all the onboard laboratories will have been running for a few years. In fact, the space research technologists will be anxious to continue inventing new concepts and refining old ones. But what happens after 2016? NASA will not longer have the money to fund ISS. Where will the next generation of researchers go?

A commercial space station is the only place they can go. The task at hand is to make it accessible. One of the most important requirements is to allow for both a manned and automated presence. Though long debated, both manned and unmanned spaceflight are necessary. The human brain is still the best computer in the solar system, but machines are much cheaper and are disposable. The main advantage to humans in space is their ability to quickly react to opportunities or crises, including the repair of defective spacecraft.

The best example of the need for humans in space was the Hubble Space Telescope repair mission. Back in 1990, astronomers were astonished by the fact that their images from the telescope were coming back blurred. Upon closer investigation, it turned out that the telescope had the wrong curvature on its main mirror. In 1993, a manned Shuttle mission was dispatched to correct the mirror with new optics. Only humans could have made such a repair in time, and if they had not, the unmanned satellite telescope would have been faulty for years.

A human presence is also necessary for a space laboratory environment. To do the best research, it is necessary to be immersed in the R&D setting. I spent many years working in engineering laboratories where polymers and pharmaceuticals were being developed. By being in that daily environment, I was inspired to invent experiments that would be inconceivable from a remote location. Inspiration cannot be scheduled. You have to be engaged in the laboratory setting to invent new experiments as you go along. Would Thomas Edison ever have forged the Industrial Revolution without living in his laboratory day and night until he found just the right materials to make the first light bulb filaments?

The advantage to machines in space is that you get more science for the buck, and this can mean the difference between a mission and no mission at all. When it comes to government or commercial funding, unmanned space projects are the darlings of frugal politicians, scientists and business people. The cost is a hundred times less to send satellites to other planets rather than humans. Scientists love them because they can satisfy narrow science goals, and record everything very accurately. The exploration of the planets by Mariner, Venera, Viking, Voyager, Magellan, Galileo, Eros, Hyubasa, or Cassini would never have taken place if a manned presence where required. The broadcasting industry would be years behind where is it now if they needed to pilot communication satellites.

But the public always felt that they were not really participating in space missions unless they could view them through the eyes of the astronaut. It never really seemed like a true space adventure unless it was experienced through the human soul.

It is important to remember that when people are sent into space, automated equipment is sent with them. Manned missions will always include probes and other instrumentation that will remain in space after the astronaut has gone. Every astronaut who ventured to the Moon did so with battery of devices that were left on the surface to continue science studies. Most notably was ALSEP, the Apollo Lunar Surface Experiments Package. This array of instruments measured the lunar heat flow, seismic activity, and radiation on the lunar surface for years after the astronauts were gone.

Not only did the Apollo astronauts solve technical glitches every day, but also retrieved large amounts of specimens. They returned with samples that later clinched the current theories on the origin of the Moon. The rocks recovered matched a theory that the Moon was chipped off the Earth by a massive foreign object. What better science could come out of manned space flight? Unmanned missions mean instrumentation is deployed; manned missions mean *both* humans and instrumentation are deployed. Commercial missions to near Earth orbit will require automated machines. And it will require qualified people to reprogram and repair them.

Commercially relevant experiments have been developing for some time on civil space stations. Tantalizing clues to plausible industrial applications began to surface in the Soviet Salute 4 and the American Skylab missions.

The first Salute space station was launched in September 1971 with little fanfare in America or Russia. At this time, the triumphant Apollo missions were at their zenith, and American astronauts were exploring the mountains of the Moon. But the USSR space program remained focused on more pragmatic goals. With its military objectives in check, the USSR was determined to create a facility in space to enhance the standard of living for its population. An array of semiconductor crystals, exotic alloys, and botanical marvels were created on the Salyuts, and these accomplishments set the standard for research today.

However, you would never have known this from the Cold War media reports at the time. Most newspapers were thundering away with heavy-handed rhetoric so biased that you would think Ming the Merciless himself was in orbit. The fact was that the Salyut space missions and the short lived American Skylab, cut our teeth on long duration space flight. Both these projects proved, though empirically, that the best commercial research takes place with long duration times and live astronauts.

Microgravity Science: The New Physics

As astronauts reach low Earth orbit they quickly notice they are getting puffy faces and upset stomachs. This is because Earth's gravity is being cancelled out by the speed of the spacecraft circling the Earth. As the vessel achieves a full orbit, the centrifugal force of the ships velocity will counteract the gravitational pull of the Earth almost exactly. There is still a little gravity; otherwise you would float away from the Earth. Vibrations from the vessel, called accelerations, will create forces too, so the weightless environment is called "microgravity" instead of "zero-gravity". In spite of this, the environment is free from most many inhibiting forces. Astronauts soon became aware that weightlessness opens up a world of phenomena I call the "The New Physics"

It is really the same old physics, but what happens in microgravity is very unique. Since there is no such thing as weight, chemicals behave differently than they do on Earth. In industry, most chemical products start in a liquid state and are then are processed into products we buy. They are usually made up of many ingredients, each of which has their own weight. Metals, plastics, glass, or colloids (milk, inks, paints are examples of colloids) and even gases all have the same weight and will interact differently in space than they do on Earth.

Because all the ingredients in a liquid formula have the same weight, they will improve their chemical performance. When you heat them up in microgravity, their molecules all move at the same rate. They will mix better, solidify better, and accumulate better, because there is no difference in the motions of each ingredient. On Earth, chemical manufacturers battle gravity by blending, agitating, freezing, smelting and solidifying without realizing it. Paint separates, metals fatigue, and glass cracks, alloys and plastics vary in their densities and quality from batch to batch. Because ingredients blend better in space, better product performance can be achieved.

Space is an especially interesting place for biotech research. In biology experiments on Earth, living cells cannot drink as much water as they want, and crystals (discussed later) grow in a haphazard manner. Artificial skin grows more modestly, and antibiotics take longer to cultivate. In space, all the processes are enhanced by the fact that the crystals or cells do not have to fight gravity was they grow.

One of the biggest markets for commercial processing in space is protein crystal growth. To explain why proteins are important, just look in the mirror. You will see proteins staring back at you: skin, eyes, hair, etc. Protein is what diseases attack in living cells. Proteins also make up viruses or bacteria that is attacking the body. By creating proteins in space, biologists can examine how they operate and perhaps figure out ways of making drug treatments and vaccines to fight diseases. Treatments for cancers are at the top of the list.

Protein crystal growth is done in fluids. The molecules have to be collected in a solution to build up the crystals for a few weeks at a time. If there are any currents or eddies in the solution due to gravitational influence (convection and sedimentation) the crystal growth formation is disrupted. In space, these convection currents are minimized and crystals grow bigger and better.

Microgravity can also be described as an ultra-low turbulence environment (there is still some turbulence due to the vibration of the space station itself). This can give scientists new perspectives on how formulas can work. Since all medical and commercial products are made of molecules of metals, plastics (polymers) or glass (ceramics), etc. a vast array of possibilities for making better products emerge, like vaccines, purified and versatile alloys, metallic glass, optical fibers, super filters, and improved solar power cells, all of which have been done on Shuttle flights.

Space also is an environment where very low contamination is possible without the elaborate vacuum systems or air filters needed on Earth. Because space is mostly a vacuum anyway, products can be made without any interference from air or dust. Simply by opening up a test chamber to space, scientists can experiment in the production of extremely clean microelectronic materials without a lot of equipment.

Space is also very hot and very cold. This change is very sudden. As the station goes from day to night, the sun plunges below the horizon. Certain crystallization applications can be devised by exposing chemicals to this rapid flash freezing or heating. This can result in better molecular structures of metal, plastic, glass that can then be used as a feedstock for other productions produced on Earth or in space. Materials can be strengthened and seasoned for long space missions or stressful applications on Earth.

Since there is no up and down in space, products can be manufactured while "levitating" in the weightless environment. If chemists are creating alloys, they do not need a vessel in which to mix the ingredients. All they have to do is keep the alloy floating in one place. This can be done with electric currents or sound waves. Containerless manufacturing allows chemists to create extremely unique compounds that will not react with the side of the vessel and become contaminated. Absolutely pure alloys can be created that could enhance or even revolutionize airplane engine parts and airframes.

Containerless manufacturing has another advantage. Spheres of metal, glass, or plastics can be manufactured that are *perfectly round*. Ball bearings, calibration beads, and microcapsules can be created with precision. In pharmaceutical applications, drugs can be formulated into tiny spheres that deliver time-release treatments to patients.

There are drawbacks to microgravity chemistry, however. Bubbles that normally come out of heated mixtures will stay in, making certain materials weak and inconsistent. Welded parts become brittle, dangerous chemicals float freely around the lab contaminating as they go, bacterial fungus grows wildly in the weightless environment, and flame becomes unpredictable and dangerous. Nothing settles down, everything floats away, and much experience is needed in creating a successful laboratory process that is commercial-friendly.

Benefits of Microgravity for Material Science Research

Effect	Benefit
Microgravity for Materials	Low molecular turbulence for the creation of products with unique qualities in shape, form, strength and performance.

Microcontamination Control	Vacuum levels can be much lower in space than on Earth leading for highly pure products.
Solar Radiation Exposure	Unfiltered ultraviolet and infrared rays help in the creation of unique products.
Flash Freezing/ Burning	Abrupt changes in temperature aid in the creation of new products such as composites.
Ionized oxygen atoms exposure.	These reactive atoms can assist in changing the properties of products.
Endurance Testing	Collectively, this harsh environment leads to benchmarking the durability of products used on Earth and in space.

What is missing in space today are industrial scientists, people who have an eye for the marketplace and know how to make things cheaply so that they can sell them for the highest profit. Through this "Edisonian perspective", innovated development can take place, products and services we simply cannot imagine will become commonplace. For the past forty years, pilots, academics, medical doctors, and a few tourists have flown in space. Their missions have been irregular and over-extended due to their high cost. With a stable, self-sustaining station in space, industrial scientist can fully utilize the microgravity environment for business research over long periods.

Commercialism in space has been going for along time, but it is still in the research phases. Pilot manufacturing never took hold. But the early discoveries in Salyut, Skylab, MIR and the Space Shuttle are the basis for the research now taking place on the International Space Station. They also serve as an analogy to turnkey commercial space facilities and how they may operate in the initial stages.

The Salyut Space Stations (1971-1986)

Salyut 1 was launched from the steppes of Kazakhstan on April 19, 1971. Considered at first to be military space station, it performed experiments that ranged from Earth observations to shooting down a satellite with a 30mm cannon! Greenhouse experiments were first conducted on Salyut 1. Cosmonauts had their first experience with extended periods of weightlessness, and the first detailed medical study of the effects on the human body in microgravity was conducted. The prognosis was grim, but the knowledge of how the body adapts to space was first recorded. Three cosmonauts were aboard Salyut for one whole month, but the mission ended tragically when they suffocated while returning to Earth.

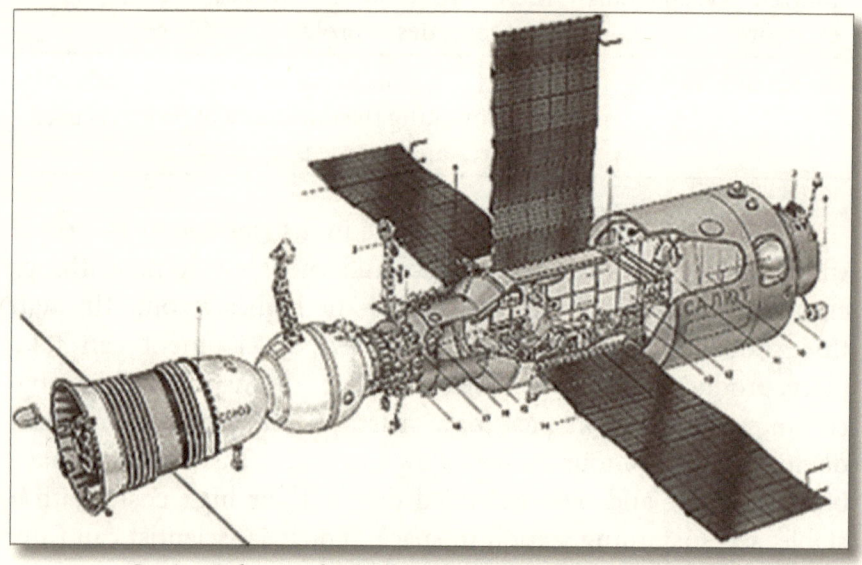

Soviet Salyut 4 docked with manned Soyuz vehicle
(Image courtesy of NASA)

Unlike the space shuttles, the Salyuts were scuttled after a few visits. They were built, deployed and decommissioned every few years. On Saluts, 3, 4, and 5 intense Earth observations continued. Salyut 5 was the first Russian material science lab. Cosmonauts began processing the first alloys, and studying the effects of microgravity of future food supplies such as fish, turtles and mushrooms. Pharmaceutical research was performed to develop ways of treating illnesses and developing vaccines. Greenhouse tests in growing vegetables from seed to seed were performed. Each mission was built on successes of the past to find the best ways to live in space for long periods.

The next space stations, Salyuts, 6 and 7, grew progressively more scientific. Many lessons on how to grow food, create better alloys, and reduce calcium depletion in bones were learned. The alloy furnaces on board created semiconductor crystals of high uniformity. Experiments on plants, such as Arabidopsis, azola, and crepis under microgravity and ambient radiation conditions were performed. But from an even wider perspective, the Soviet space program first taught the world just what habitable space was like.

Salyut 6 and 7 vessels showed an evolution in design. These stations allowed for two docking ports on either end, instead of only one. This enabled the Soyuz launch vehicle to dock on either port, leaving an open port for its sister spacecraft, the Progress cargo carrier, to dock with supplies. These two vehicles were a fraction of the cost of Space Shuttle, and could fly with or without crews. The Progresses could re-supply the stations for very long stays in space, culminating in a record residence times that American astronauts could not approach.

Internationalism was also a priority for the Soviets. Starting with Salyut 6, the Soviet space agency created Intercosmos, an effort to bring cosmonauts from other countries aboard the Salyuts. Scientists from Brazil, Cuba, France, India, and Viet Nam all worked on science efforts to help alleviate their country's dire economic conditions with new models for research. Thought to be a Cold War publicity stunt by the west, the Soviets sponsored international cosmonauts who would normally never be able to afford the trip. The Salyuts were the first international space stations.

Salyut 7, launched in 1982, was the most evolved of all. By using modules to build up the station with successive segments, it set the stage for the next version, the space station MIR. It housed six crews is its lifetime, including a whopping 237 day expedition. It was ready for decommission by 1986 and was scuttled in 1991.

Skylab: A Glimpse into the Industrial Potential of Space

Inconsistency in the American space program began right after the Apollo missions. The 1970s was a period of transition that I call "The Great Recession". From 1973-1984, three recessions took place, broken up by periods of tentative improvement. Presidents, Richard Nixon, Gerald Ford, and Jimmy Carter wrestled with Watergate, Viet Nam, and relentless inflation. Unemployment was at the highest levels since the 1930s, inflation at the highest level since World War II. The public seemed to become concerned less about the space program than about ending the war, saving Earth's environment, and alleviating joblessness.

This is where American congressional leadership should have stood up and stayed on course with manned space exploration. Instead, they perceived that the nation as "bored" with space and the economic challenges did not warrant large space projects NASA envisioned. Anyone who lived in those turbulent days would remember how politically unfeasible large space stations or Moon missions were. Ubiquitous economic strains made these quests untenable in the post Viet Nam era

Most space historians believe the fatal blow actually came in the late sixties when Congress cancelled the Saturn Moon rockets in favor of the Space Shuttle. The Saturns were essential to creating an infrastructure to the Moon and the construction of large space stations. This decision delayed space commercialization for 10 years, and the return to the Moon for 50 years. If Congress had improved Saturn boosters instead of canceling them, the rockets could have been used to build and re-supply a space station and even a Moon base. We would be planning a manned Mars landing right now instead of having to start from scratch with Project Constellation.

The 1970s "new vision" was one of efficiency and practical necessity, not of deep space exploration. NASA had to use up old hardware, and deflect public criticism emphasizing the industrial and environmental potential of a laboratory in space. The Skylab program fit neatly into this new era.

Richard Nixon released this statement about Skylab after the Saturn boosters were cancelled:

> (Skylab) will help develop new methods of learning about the Earth's resources, and new methods of evaluating programs aimed at preserving and enhancing the resources of the entire world. It will

seek new knowledge about our own star, the sun, and about its tremendous influence on out environment. Scientists about the Skylab will perform medical experiments aimed at a better knowledge of man's on physiology. Also, they will perform experiments aimed at developing new industrial processes utilizing the unique capabilities found in space.

The potential commercial applications discovered on Skylab would be significant. But first NASA had to tackle the question of whether to have people in space at all. Do you need humans to conduct research? Why not use an unmanned facility that can be operated from the ground? As it turned out, you absolutely *need* humans in space.

The Skylab Space Station 1973-1974 *(Image courtesy of NASA)*

Skylab was launched in May 14, 1973 aboard the last built Saturn V Moon rocket. The facility itself consisted of an empty Saturn S-

IVB third stage fitted with all the equipment for an extended stay in space. Werhner von Braun, an early advocate of the orbiting wheel space station design depicted in 2001: A Space Odyssey, suggested using an even bigger fuselage for Skylab, but problems of getting it into orbit dictated that Skylab stay small and inexpensive. The staff at NASA's Apollo Applications Program thought up a thorough mission, filled with promise and optimism, but near tragedy struck in the first phase.

Sixty-three seconds after liftoff, the meteoroid shield, which was also designed to shade Skylab from the sun, ripped loose. It disturbed the mounting of one of the workshop's solar panels causing it to only partially deploy. The exhaust plume of the second stage retro-rockets then ripped away the other solar array. There was no heat shielding left on the spacecraft either. Skylab started dying the day it was born.

Without the solar arrays providing power, the station could not stay in a stable orbit or keep itself cool from the merciless solar rays. It was spiraling down and getting hotter and hotter inside. The first crew had to perform a rescue mission, and there was no time for training.

What transpired is another one of the clearest examples of why we need people in space. Within eight days the first crew was on its way and trained to handle the problems. The initial crew, consisting of Charles "Pete" Conrad, Paul J. Weitz, and Joseph P. Kerwin, spent twenty-eight days aboard Skylab. They went EVA to free up the solar panel and placed a makeshift parasol over the workshop's outer hull to provide the much needed shade. This saved the entire mission from disaster. It showed that quick responses to crises and opportunities are endemic to space exploration and utilization where mid-course adjustments are necessary and expected.

The material science experiments performed aboard Skylab were very diverse. But of the 171-days Skylab operated, a mere 30 hours was devoted to them. Three crews performed tests in alloy production, semiconductor fabrication, combustion efficiency, and composites production. In these long duration facilities, the ability to analyze and adjust conditions on-orbit was critical to the continuity and micro-transitions in experimentation. If you can change the experiments as you do them, a surprising array of phenomena could emerge in a very short period.

Skylab's Commercially Relevant Experiment Results
(Courtesy of NASA)

Exp. #	Description	Results
M479	Zero Gravity Flammability. Determine degradation effects of pre-launch, launch, and space environments on absorptivity/emissivity characteristics of thermal control coatings.	Data on toxicity, contamination, cleaning, timeline, and hardware performance were provided.
M512	Materials Processing Facility[6]. Explore space manufacturing applications of molten phenomena, such as molten metal flow, freezing patterns, thermal stirring, and fusion across gaps, and surface tension, by performing five experimental tasks and Experiment 479.	The facility and the returned samples were identical to the training hardware and samples. The welding went extremely well. The materials processing facility was used to conduct experiments M 479, M 551, M 552, M 553, and M 555.
(M551)	Metals Melting. Study the behavior of molten metals in microgravity. Characterize the structures formed in metals melted and rapidly solidified in zero gravity. Test means joining metals by electron beam welding in zero gravity.	Ground specimens contained large elongated grains and a wide chill zone. The Skylab specimen contained more equiaxed fine grains and a more symmetrical pattern of grain structure. The finer grain structure observed in the space specimen was attributed to constitutional supercooling, which results when a solid freezes with a composition slightly different from that of the liquid from which it forms.

(M552)	Exothermic Brazing. Test and demonstrate a method of brazing components in space repair and maintenance operations. Study surface wetting and capillary flow effects in weightless molten metals.	In the same time and at the same temperature conditions, nickel dissolved more rapidly in liquid silver copper alloys in space than on Earth. The experiment indicated that this occurred not because the nickel was more soluble in space, but because the speed of dissolution was greater. This suggests that saturated liquid metal solutions can be more easily produced and true solubility more easily determined in space than on Earth.
(M553)	Sphere Forming. Demonstrate the effects of zero gravity on fundamental solidification phenomena.	Gun problems resulted in the specimen being pear shaped instead of spherical following gun cutoff. Incomplete melting occurred, and sometimes the spike would retract and the sphere would stick on the ceramic. At about the same time the gun problems were occurring, the time required to create a vacuum in the system seemed tremendously long. Apparently, outgassing in the gun was causing the vacuum problem.
(M555)	Gallium Arsenide Crystal Growth. Grow single crystals of gallium arsenide from solution in order to produce material of exceptionally high chemical and crystalline perfection.	The experiment was successfully carried out. Knowledge of the role of gravity in materials processing has made substantial progress.

M518	Multipurpose Electric Furnace System. Enhance the capabilities of existing Skylab hardware by providing means to perform experiments on solidification, crystal growth, and other processes involving phase changes in materials.	The furnace system performed well and no malfunctions were encountered. Experiments M 556, M 557, M 558, M 559, M 560, M 561, M 562, M 563, M 564, M 565, and M 566 were performed using the multipurpose electric furnace system or the material processing facility. All samples processed in the furnace were returned to Earth. Results from some of the processes were far superior to the results obtained on Earth.
(M556)	Vapor Growth of II-VI Compounds. Determine the degree of improvement that can be obtained in the perfection and chemical homogeneity of crystals grown by chemical vapor transport under weightless conditions in space.	Mixed crystals of compound, semiconductor germanium selenide and germanium telluride were grown by chemical transport through a temperature gradient in a transport agent, iodine vapor, from polycrystalline sources of the two component materials. The growth process was carried out in sealed quartz ampoules contained in the sample cartridges.
(M557)	Immiscible Alloy Compositions. Determine the effects of near zero g on the processing of material composition that normally segregate on Earth.	It was demonstrated that a completely stable dispersion of the two immiscible liquids, which were very unstable on Earth, can be prepared in space. Since the important parameters of immiscible liquids, such as viscosity and density differences, are similar to those found for common liquid metal immiscible systems, the metallic systems should also be very stable in low gravity.

(M558)	Radioactive Tracer Diffusion. Measure self-diffusion and impurity diffusion effects in liquid metals in space flight, and characterize the disturbing effects, if any, due to spacecraft accelerations.	A marked decrease in zinc 65 movements along the length of a cylinder of liquid zinc in space was apparently caused by the absence of convective mixing. The radial distribution observed in the Skylab samples also indicated that convective mixing was negligible in space.
(M559)	Microsegregation in Germanium. Determine the degree of microsegregation of doping impurities in germanium caused by convectionless directional solidification under conditions of weightlessness.	Space grown crystals were compared with identical crystals resolidified on Earth. Microsegregation in space is one half to one fifth that on Earth in the bulk material which implies a reduced diffusion or mass transport of the solute through the host material during solidification
(M560)	Growth of Spherical Crystals. Grow doped germanium crystals of high chemical homogeneity and structural perfection and study their resulting physical properties in comparison with theoretical values for ideal crystals.	Single crystals with extremely low density of defects were obtained. Even though the crystals were small, very large crystals could be prepared by this approach. The technique would seem to be ideal for processing of highly reactive and high melting temperature materials. Since no mechanical feedthroughs are required, the technique could be most readily adapted to high pressure or encapsulated growth.

(M561)	Whisker-Reinforced Composites. Produce void free samples of silver or aluminum, reinforced with oriented silicon carbide whiskers.	The experiment produced void free samples of silver, reinforced with oriented silicon carbide whiskers. Sintered rods of silver containing distributions of unidirectionally oriented silicon carbide whiskers, one micron in diameter by one millimeter long, were melted in the furnace. Pressure was exerted to force voids from the melt and promote wetting of the whiskers by the matrix material.
(M562)	Indium Antimonide Crystals. Produce doped semiconductor crystals of high chemical homogeneity and structural perfection and to evaluate the influence of weightlessness in attaining these properties.	High quality single crystals of indium antimonide, doped with tellerium, were precision machined and etched to fit into heavy quartz ampoules, sealed, and enclosed in metal cartridges. Half of each crystal (7.62 cm in length) was melted in the furnace and regrown at the rate of 1.27 cm per hr using the unmelted half as seed.
(M563)	Mixed III-V Crystal Growth. Determine how weightlessness affects directional solidification of binary semi conductor alloys and, if single crystals are obtained, determine how their semiconducting properties depend on alloy composition.	Alloys of indium antimonide and gallium antimonide in varying proportions were placed in separate, fused silica ampoules, encased in cartridges, melted in the furnace, and directionally solidified at the slowest available rate.
(M564)	Halide Eutectics. Produce highly continuous, controlled structures in samples of the fiberlike NaF NaCI and platelike Bi Cd and Pb Sn eutectics, and measure their physical properties.	The experiment produced controlled structures in samples of fiberlike, fluoridesodium chloride eutectic, and measured their physical properties. Three ingots of the eutectic, 1.27 cm in diameter and 10.16 cm long, were grown by melting the alloys and then cooling them directionally at the slowest available rate.

(M5B5)	Silver Grids Melted in Space. Determine how pore sizes and pore shapes change in grids of fine silver wires when they are melted and resolidified in space.	The action of diffusion and of the remaining convection due to the variations in [394] the surface tensions appeared to be reduced in space from the rapid leveling of concentration gradients on Earth experiments.
(M566)	Copper-Aluminum Eutectic. Determine the effects of weightlessness on the formation of lamellar structure in eutectic alloys when directionally solidified.	Three aluminum copper alloy rods 0.64 cm in diameter were partially melted and directionally solidified.

The high cost, short residence times and diversified science manifest that characterized the Skylab program, pre-empted the development of these technologies. Because it was all so new, no killer application evolved; no transforming breakthrough emerged to cure diseases or create new energy sources. Yet the hope remained that these little known beginnings in material science might culminate into billions of dollars in revenue for the next generation of space entrepreneurs.

By 1974, Skylab was abandoned and went on to circle the Earth as a ghost ship. No vehicle existed to service it. Escaping from the labyrinth of recession and political disillusionment, Skylab flew, only to fall short of its potential.

Apollo-Soyuz Test Project – Da, the Science Too

Despite the political turbulence of the mid-seventies, the Soviet Union and the United States could agree on a *few* things. One of them was the need for rescue capability in space. Both countries had common equipment in orbit, so why couldn't they join spaceships in a test flight to see if we could help each other out in an emergency. This required a compatible docking mechanism that would link the two cold worlds together. The Apollo-Soyuz Test Project's was launched in 1975. On its 9-day mission, the world watched one of the few demonstrations of détente, and ignored any science that was performed. Yet a common thread was emerging that became stronger over the years: space internationalism.

Not just détente. The Apollo Soyuz Test Project (ASTP) included material science payloads and radiation tests on plant seeds, shrimp eggs, and bacteria. *(Image courtesy of NASA)*

Overtures of space cooperation between the superpowers had been going on since the days of Sputnik The Kennedy administration, flushed from their success of the John Glen Mercury flight, established certain projects that would create a common benefit for both the Soviet Union and the U.S. As summarized in 1962 by Arnold Frutkin, NASA's Director of the Office of International Cooperation, in his letter to John F. Kennedy, he recommended.

1. The establishment of an operational world weather satellite system through the coordinated launching by the US and the USSR of weather satellites in complementary orbits, the resulting data to be distributed globally through existing meteorological channels;
2. The exchange of spacecraft tracking services, each side providing equipment suited to its own requirements to be

erected and operated on the other's territory by the other's technicians;

3. Mapping of the earth's magnetic field in space, a matter "central to many scientific problems," by satellites which the countries would launch, one each, in complementary orbits;

4. An invitation to the Soviet Union to join in programs already under way with other countries for the joint testing of intercontinental communications satellites (each country providing a ground terminal suitable for working with US communications satellites and participating in an international ground station coordinating committee).

Space cooperation, even with its difficulties, seemed more affordable with unilateral efforts, especially in light of the alternatives. President Kennedy was receptive to the idea, and in 1963, in a speech before the 18th General Assembly of the United Nations he made a proposal for a joint expedition to the Moon!

> Finally, in a field where the United States and the Soviet Union have a special capacity-in the field of space-- there is room for new cooperation, for further joint efforts in the regulation and exploration of space. I include among these possibilities a joint expedition to the moon. Space offers no problems of sovereignty; by resolution of this Assembly, the members of the United Nations have foresworn any claim to territorial rights in outer space or on celestial bodies, and declared that international law and the United Nations Charter will apply. Why, therefore, should man's first flight to the moon be a matter of national competition? Why should the United States and the Soviet Union, in preparing for such expeditions, become involved in immense duplications of research, construction, and expenditure? Surely we should explore whether the scientists and astronauts of our two countries-- indeed of all the world--cannot work together in the conquest of space, sending someday in this decade to

the moon not the representatives of a single nation,
but the representatives of all of our countries.

As the Cold War propaganda raged in the press through the 1960s, the leaders of both countries contemplated the idea of peaceful cooperation in space, at least in principal. In 1972, the improbable happened. Richard Nixon and Soviet Prime Minister, Aleksey Kosygin signed an agreement to cooperate in space and to join together two spacecraft in low earth orbit. The Apollo-Soyuz Test Project was born.

The docking of the Apollo and Soyuz capsules took place on July 17, 1975. Alexi Leonov, Valeriv Kubasov, Thomas Stafford, Vance Brand, and Donald "Deke" Slayton joined together in the continuing international effort to colonize space: a remarkable adventure in a precarious time in history. Soviet aggression was again on the march throughout Asia and Latin America. Nobel Prize winner, Alexander Solzhenitsyn had recently published his expose' "The Gulag Archipelago" on Soviet human rights atrocities in Siberian prison camps. Relations would soon get chillier in the years to come, but at that time, it still seemed like the leaders of both countries were really trying to reduce the general tensions of the Cold War. A break in the clouds of a forty-year storm.

The docking lasted only two days, and other than being a footnote in the history of manned space flight, nothing much was heard thereafter. Apollo-Soyuz was, in fact, the template of every future space station flight on either side of the Iron Curtain.

Apollo-Soyuz Application Development Manifest

MA-011 Electrophoresis	Isolation of urokinase enzymes, for helping researches to find drugs to treat heart attacks, clots and blood diseases.
MA-014 Electrophoresis Enhancements	A German experiment to enhance space-based and ground bases isolation of biomedical chemicals.

MA-010 Multipurpose Electric Furnace Experiment System	Material science processing experiments involving phase changes of solids, liquids and vapors at elevated temperatures.
MA-041 Surface Tension Induced Convection	This experiment was to study the effects of surface tension on molten metals in a weightless environment.
MA-044 Monotectic and Syntectic Alloys	Blending aluminum antimony alloy for better solar cells.
MA-060 Interface Marking of Crystals	Analysis of germanium crystal growth for better semiconductors
MA-070 Processing of Magnets	Blending of magnetic metals for even melt and betting EM properties.
MA-085 Crystal Growth from Vapor Phase	Making pure and symmetrical semiconductor.
MA- 131 Halide Utectics	Fiber optics research
MA-150 U.S.S.R. Multiple Material Melting	Melting and re-solidifying metals to make them more uniform in microgravity.
MA-028 Crystal Growth	The effects of microgravity on a water based crystal-forming solution for insights into better semiconductor fabrication.

The Cold War winds then changed direction. The Soviet Union invaded Afghanistan four years later, which, as most historians agree, eventually precipitated the September 11, 2001 attacks on New York City and Washington, DC. But years of international influence permeated the Salyut, MIR, Shuttle and ISS programs even though the media, and our leaders, east and west, continued to fan the flames of political dissent. Compared to the Moon race, space station research did not seem technically enlightened or politically progressive. It lacked the sexiness of a lunar base, but gave both the Soviet Union and the U.S. a common high ground to forge a new beginning. It dawned on both nations that peace-fostering trade could be done in space as well as on Earth.

The Shuttle Era's Commercial Accomplishments

In the 1980s, America emerged from "The Great Recession" a much different nation. Politically, a conservative wave, struggling to emerge from the incredulity of the post-Nixon years, broke free, after the Jimmy Carter's regime failed to resurrect the faltering American economy. The hard times dragged on into the early eighties. It wasn't until Ronald Reagan's second term before the U.S. was back on a prosperous course.

As the Soviets slowly evolved space applications on the Salyut stations, the Space Shuttle flew in fits and starts. The reusable spacecraft was to be the baseline for the exploration of the rest of the solar system. But could it close the gap between the U.S. and the Soviets? By the time the first Shuttle was launched in 1981, the Soviets had been in space almost continuously for ten years. The U.S. "return to space" looked more like a retrograde course back into the early sixties.

The Shuttle's commercial agenda was bold. It was to be a space laboratory, launch platform, satellite repair shop, and cargo carrier for building a long duration space station. It was to fulfill every requirement the American space program demanded with the least money possible. This meant that commercial applications were not only expected, but were essential for the survivability and credibility of the manned space program goals. The primary mission of the Shuttle was to build a space station that would provide long-term habitability. In the meantime, it was to provide a platform for Spacelab, a short duration space applications laboratory. Since applications take a long time to develop in space, a short-term space experimental facility seemed oxy-moronic.

The core of the commercial Shuttle program would be the European Spacelab module. Over its 17-year history, many of these missions were dedicated to material science, some of which were commercially intended. Europe's goals, like those of China, were devoted to improve social and economic development on Earth using products and processes developed in microgravity. It was believed that space exploration fostered these innovated technologies. Over 750 experiments in almost every discipline were performed in the

Shuttle's Spacelabs, but they only set the precedents for what could be accomplished in a long-term facility.

According to NASA, the Shuttle had 77 "commercial flights" with 266 experiments on board between 1982 and 2000. They ranged from protein crystal growth to continuous electrophoresis (a chemical separation and isolation process). Others were semiconductor manufacturing (Wake Shield Facility), generic bioprocessing, gas permeable polymers, zeolite crystal growth and space exposure tests on various materials. The most often flown process was protein crystal growth experiments that were on about 36 commercial flights within that time frame.

Many companies were involved in shuttle experiments, such as 3M, Bristol Myers Squibb, Schering Plough, Amgen, International Flavors and Fragrances and Polyfoam. Among the products developed were proteins to study diseases, latex spheres for the calibration of equipment, perfume fragrances, and Aerogel, which is an ultra low-density insulating material. One of the latest products was a permanent contact lens that need not be taken out of the eye because it could "breathe" enough to be comfortable. Many companies were interested in continuing on the International Space Station before the Vision For Space Exploration cut back commercial science in favor of developing a lunar program.

The Shuttle commercial flight model was significant because it showed that scientists and engineers could go into space for a short periods of time, do experiments efficiently, and return to Earth in a business-like fashion, without extensive adaptation to space.

Commercial Flights and Experiments of the Shuttle Program 1982-2000

Date	Flight	Payload Name
March 22, 1982	STS-03	Monodisperse Latex Reactor System.

June 27, 1982	STS-04	Monodisperse Latex Reactor System. Continuous Flow Electrophoresis System.
April 04, 1983	STS-06	Continuous Flow Electrophoresis System. Monodisperse Latex Reactor System
June 18, 1983	STS-07	Continuous Flow Electrophoresis System Monodisperse Latex Reactor System. Single Axis Acoustic Levitator.
August 30, 1983	STS-08	Continuous Flow Electrophoresis System
February 03, 1984	STS-10	Monodisperse Latex Reactor System.
April 06, 1984	STS-11	Solute Diffusion Appartus.
August 30, 1984	STS-12	Continuous Flow Electrophoresis System.
Nov. 8 1984	STS-14	Diffuse Mixing of Organic Solutions
April 12, 1985	STS-16	Continuous Flow Electrophoresis System.
June 17 1985	STS-18	Protein Crystal Growth
July 29 1985	STS-19	Protein Crystal Growth
August 27 1985	STS-20	Physical Vapor Transport of Organic Solids
October 30 1985	STS-22Sing	Single Axis Acoustic Levitator.
November 26 1985	STS-23	Continuous Flow Electrophoresis System. Diffuse mixing of Organic Solutions. Protein Crystal Growth.
January 12, 1986	STS-24	Protein Crystal Growth.
CHALLENGER	**DISASTER**	**January 1986**

September 29, 1988	STS-26	Physical Vapor Transport of Organic Solids. Protein Crystal Growth.
March 13, 1988	STS-29	Protein Crystal Growth.
March 29, 1989	Consort-01	Three Dimensional Microgravity Accelerometer. Electrophoresis Cells. Elastomer-Modified Epoxy Resins. Foam Formation Device. Immiscible Polymers. Liquid Phase Sintering Furnace. Materials Dispersion Apparatus.
May 4, 1989	STS-30	Fluid Experiment Apparatus
October 18, 1989	STS-34	Polymer Morphology
November 15, 1989	Consort-02	Three Dimensional Microgravity Accelerometer. Automated Generic Bioprocessing Apparatus. Biomodule (PSB). Electrodeposition cells. Elastomer-Modified Epoxy Resins. Foam Formation Device. Immiscible Polymers. Investigations into Polymer Membrane Processing. Liquid Phase Sintering Furnace. Materials Dispersion Apparatus. Polymer Curing Experiment. Plasma Particle Generation. Thin Films.
January 9, 1990	STS-32	Fluid Experiment Apparatus. Protein Crystal Growth.
April 24, 1990	STS-31	Investigations into Polymer Membrane Processing. Protein Crystal Growth.

May 16, 1990	Consort-03	Three Dimensional Microgravity Accelerometer. Automated Generic Bioprocessing Apparatus. Biomodule (PSB). Electrodeposition cells. Elastomer-Modified Epoxy Resins. Foam Formation Device. Immiscible Polymers. Investigations into Polymer Membrane Processing. Materials Dispersion Apparatus. Nuclear Track Detector. Polymer Curing Experiment. Plasma Particle Generation. Thin Films.
October 6, 1990	STS-41	Investigations into Polymer Membrane Processing. Physiological Systems Experiments.
December 2, 1990	STS-35	Atomic Oxygen.
April 5, 1991	STS-37	Bioserve Instrumentation Materials Dispersion Apparatus. Protein Crystal Growth.
June 5, 1991	STS-40	Immiscible Polymers. Electrophoresis cells. Non-Linear Optical Material. Nuclear Track Detector.

June 18, 1991	Joust -01	Three Dimensional Microgravity Accelerometer. Automated Generic Bioprocessing Apparatus. Biomodule (PSB). Electrodeposition cells. Foam Formation Device. Investigations into Polymer Membrane Processing. Material Dispersion Apparatus. Polymer Composite Curing Experiment. Powdered Materials Processing Experiment. Plasma Particle Generation. Thin Films.
August 2, 1991	STS-43	Bioserve Instrumentation Materials Dispersion Apparatus. Investigations into Polymer Membrane Processing. Protein Crystal Growth.
September 12,1991	STS-48	Bioserve Instrumentation Materials Dispersion Apparatus. Protein Crystal Growth.
November 16, 1991	Consort -04	Three Dimensional Microgravity Accelerometer. Automated Generic Bioprocessing Apparatus. Biomodule (PSB). Electrodeposition cells. Equipment for Controlled Liquid Phase Sintering Exp. Investigations into Polymer Membrane Processing. Material Dispersion Apparatus. Polymer Curing Experiment. Polymeric Foam Formation. Space Formed Structural Beam.

January 22, 1992	STS-42	Gelation of SOLS: Applied Microgravity Research. Investigations into Polymer Membrane Processing. Protein Crystal Growth.
March 24, 1992	STS-45	Investigations into Polymer Membrane Processing.
May 7, 1992	STS-49	Protein Crystal Growth.
June 25, 1992	STS-50 USML-1	Astroculture. Commercial Generic Bioprocessing Apparatus. Directed Polymerization Apparatus. Investigations into Polymer Membrane Processing. Protein Crystal Growth. Zeolite Crystal Growth.
July 31, 1992	STS-46	Three Dimensional Microgravity Accelerometer. Electrodeposition cells. Evaluation of Oxygen Interaction with Materials. High Temperature Superconductors. Limited Duration Space Candidate Materials Exposure. Pituitary Growth Hormone Cell Function.
September 10, 1992	Consort-05	Three Dimensional Microgravity Accelerometer. Biomodule (PSB). Electrodeposition cells. Equipment for Controlled Liquid Phase Sintering Exp. Material Dispersion Apparatus. Organic Separation. Performance of LEDs in Microgravity. Polymeric Foam Formation. Space Formed Structural Beam.
September 12, 1992	STS-47	Protein Crystal Growth.

October 22, 1992	STS-52	Commercial MDA ITA Experiments. Crystals by Vapor Transport Experiment. Protein Crystal Growth. Physiological Systems Experiments
January 13, 1993	STS--52	Commercial Generic Bioprocessing Apparatus
February 19, 1993	Consort-06	Three Dimensional Microgravity Accelerometer. Biomodule (PSB). Electrodeposition cells. Equipment for Controlled Liquid Phase Sintering Exp. Material Dispersion Apparatus. Organic Separation. Polymeric Foam Formation. Space Formed Structural Beam.
April 8, 1993	STS-56	Commercial MDA ITA Experiments.
June 21, 1993	STS-57 SH-01	Three Dimensional Microgravity Accelerometer. Astroculture. Bioserve Pilot Laboratory. Commercial Generic Bioprocessing Apparatus. Equipment for Controlled Liquid Phase Sintering Exp. Gas Permeable Polymeric Materials. Investigations into Polymer Membrane Processing. Liquid Encapsulated Melt Zone/ Fluid Experiment Apparatus. Non-Linear Optical Material. Organic Separation. Protein Crystal Growth. Physiological Systems Experiments. Solution Crystal Growth. Zeolite Crystal Growth.

September 12, 1993	STS-51	Investigations into Polymer Membrane Processing. Protein Crystal Growth. Limited Duration Space Candidate Materials Exposure.
February 3, 1994	STS-60 SH-02	Three Dimensional Microgravity Accelerometer. Astroculture. Bioserve Pilot Laboratory. Biomodule (PSB). Bioserve Pilot Laboratory. Commercial Generic Bioprocessing Apparatus. Containerless Process Coating. Equipment for Controlled Liquid Phase Sintering Exp. Fast Plants. Materials Testing Laboratory. Molecular beam epitaxy. Microgravity Measurement Device. Organic Separation. Protein Crystal Growth. Space Experiments Facility. Wake Shield Facility.
March 04, 1994	STS-62	Commercial Generic Bioprocessing Apparatus. Limited Duration Space Candidate Materials Exposure. Potential Crystal Growth. Physiological Systems Experiments.
April 09, 1994	STS-59	Non-Linear Optical Material.
July 08, 1994	STS-65	Potential Crystal Growth.
September 09,	STS-64	Robot Operated Materials Processing System.
September 30, 1994	STS-68	Potential Crystal Growth.
November 3, 1994	STS-66	Potential Crystal Growth.

February 03, 1995	STS-63 SH-03	Three Dimensional Microgravity Accelerometer. Astroculture. Bioserve Pilot Laboratory. Equipment for Controlled Liquid Phase Sintering Exp. Fluids Generic Bioprocessing Apparatus. Gas Permeable Polymeric Materials. Protein Crystal Growth.
March 2, 1995	STS-67	Commercial MDA ITA Experiments. Protein Crystal Growth.
July 13, 1995	STS-70	Protein Crystal Growth.
September 7, 1995	STS-69	Commercial Generic Bioprocessing Apparatus. Commercial MDA ITA Experiments. Molecular Beam Epitaxy. Non-Linear Optical Material. Wake Shield Facility.
October 20, 1995	STS-73	Three Dimensional Microgravity Accelerometer. Astroculture. Commercial Generic Bioprocessing Apparatus. Protein Crystal Growth. Zeolite Crystal Growth.
January 11, 1996	STS-72	Protein Crystal Growth.
February 22, 1996	STS-75	Protein Crystal Growth.
March 22, 1996	STS-76/MIR	Equipment for Controlled Liquid Phase Sintering Exp.
April 3, 1996	Conquest-01	Three Dimensional Microgravity Accelerometer. Electrodeposition Cells. Foam Inflated Rigidized Structures. Materials Dispersion Apparatus. Organic Separation.

May 19, 1996	STS-77 SH-04	Advanced Separation. Commercial Float Zone Furnace. Commercial Generic Bioprocessing Apparatus. Fluids Generic Bioprocessing Apparatus. Gas Permeable Polymeric Materials. Protein Crystal Growth. Plant Generic Bioprocessing Apparatus. Space Experiments Facility.
September 16, 1996	STS-79 SH-05, S/MM-04 MIR-04	Three Dimensional Microgravity Accelerometer. Extreme Temperature Translational Furnace. Protein Crystal Growth. Commercial Generic Bioprocessing Apparatus. Materials in Devices as Superconductors.
October 23, 1996	1996 Meteor-01	Three Dimensional Microgravity Accelerometer.
November 19, 1996	STS-80 WSF-03	Commercial MDA ITA Experiments. Wake Shield Facility.
April, 4 1997	STS-83 MSL-01	Plant Generic Bioprocessing Apparatus
May 15, 1997	STS-80	Protein Crystal Growth.
July 01, 1997	STS-84 S/MM-06	Plant Generic Bioprocessing Apparatus.
September 25, 1997	STS-86-MIR-07	Protein Crystal Growth Advanced Separation. Commercial Generic Bioprocessing Apparatus. Commercial MDA ITA Experiments. Equipment for Controlled Liquid Phase Sintering Exp.

January 22, 1998	STS-89 MIR-08	Astroculture. Optimization Liquid Phase Sintering Experiment. X-Ray Detector Test.
May 28, 1998	STS-91	Protein Crystal Growth.
October 29, 1998	STS-95 Spacehab-SM	Protein Crystal Growth. Advanced Separation. Aerogel. Astroculture. Biodyn. Commercial Generic Bioprocessing Apparatus. Commercial ITA BioprocessingExperiment. Microencapsulation Electrostatic Processing System. Non-Linear Optical Material.
January 21, 1999	STS-93	Aerogel. Commercial Generic Bioprocessing Apparatus.
May 19, 2000	STS-101	Astroculture, Protein Crystal Growth.
October 31, 2000	**STS-102**	**Expedition 1 to the International Space Station**

The MIR Space Station: Prelude to Industrialization

The Russian Space Station MIR began assembly in February 1986 and stayed in space for 15 years. It turned out to be the model for civil space stations for years to come. Being "modular" in design, each segment was launched separately and had its own function. They were all attached to a main node like spokes on a hub. The concept was to have as many ports ready for attaching labs, docking supply ships and rotating crews. One such module was called the Kvant and the others, Krystall, Priroda and Spektr.

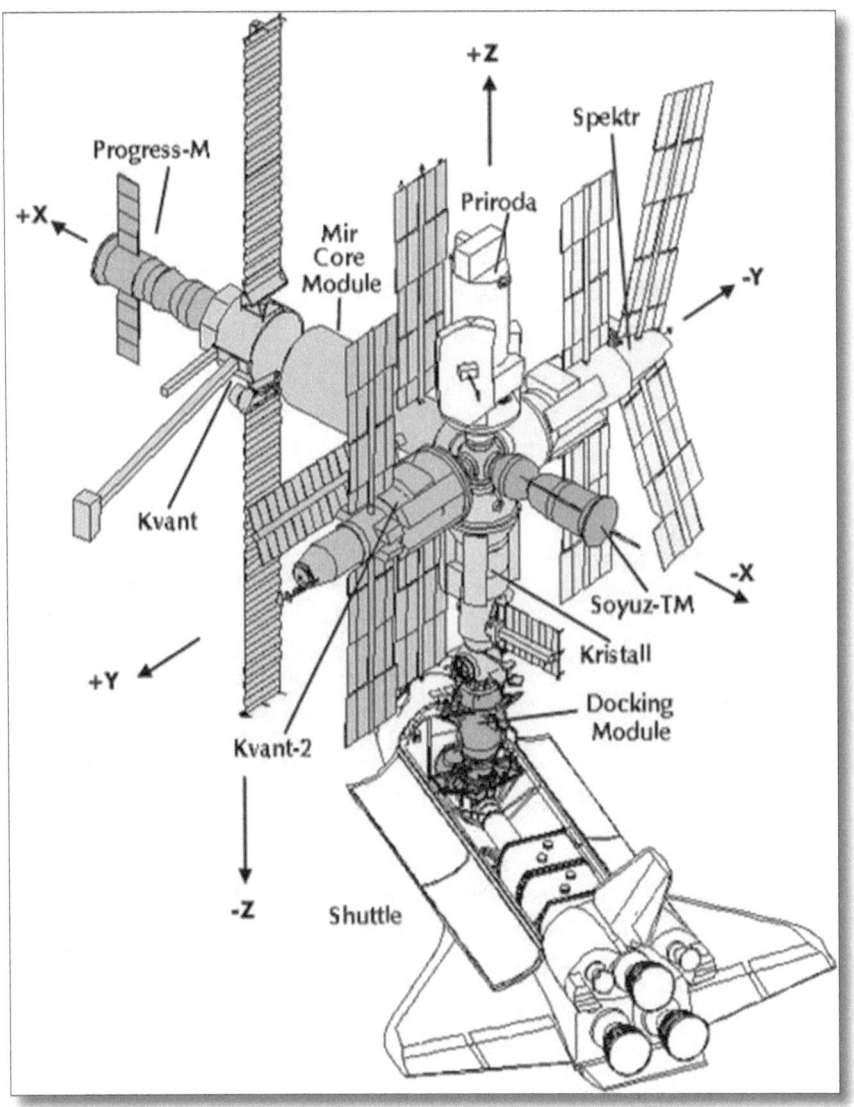

(Drawing courtesy of NASA)

The MIR space station was a boondoggle. It weighed about 150 tons, with dimensions of roughly 115 feet by 98 feet and an inner volume of about 14,100 cubic feet, significantly larger than Skylab. MIR stayed continually occupied in space for ten years. In addition to hundreds of science experiments, the facility even housed the first civilians: a Japanese journalist and a British contest winner. Each of these "visitors" stayed for about a week in space.

MIR's commercial accomplishment was the continued quest for living in space for long durations and still being productive and healthy. The results were disconcerting. Cosmonaut's lost bone mass of at least 1-2% per month. Exercise had very little effect on the bones in the legs, but kept muscle mass reasonably strong. Cosmonauts learned to truly adapt, and long duration studies culminated in a record 438-day residence time in space by Valery Polyakov in March of 1995. Radiation and demineralization effects apparently did not harm him in the long term, though no medical details were made public.

Facilities on the MIR

Module Name	Functions.
Kvant-1	Kvant-1 is the Astrophysics module, providing information for research into the physics of galaxies, quasars, and neutron stars' spectra and X-ray emissions. Kvant-1 is 19 feet long and 14 feet in diameter.
Kvant-2	Kvant-2 provided biotechnology research data, Earth observation photographic equipment, and EVA capability. It is over 40 feet long and 14 feet in diameter. An outside airlock allowed for experiments of space exposure on electronics and construction materials.
Kristall	Kristall. This technological module was used for biological and materials processing technology development. It contained equipment that produced semiconductors and other high-tech materials. Other apparatus included a greenhouse.

Spektr	The Spektr module contained equipment for Earth atmospheric research and surface studies. (Damaged and closed after collision with Progress vessel)
Priroda	A remote sensing module used for measuring ozone and aerosol concentrations in the atmosphere.

Despite frequent technical challenges, microgravity science on MIR continued in earnest. Semiconductor and superconductor alloys were blended for use in solar panels and infrared optics, protein crystals were grown for the treatment of various diseases, and wheat was grown in space from seed to seed. Electrophoresis, a process to isolate and purify human growth hormone and insulin was performed. Long duration exposures of materials were conducted outside the station. Having a stable, on-orbit laboratory was a boon to any research efforts that required undisturbed process development for weeks and month at a time. On MIR, microgravity research grew ten years beyond the levels of the Americans or Europeans.

MIR had a broad array of experimental equipment on board that spanned the spectrum of applied science. Despite the engineering difficulties in assembling and supplying a modular space station with a finicky radar docking system, experiments were continuously performed and cosmonauts rotated on schedule.

In contrast to the American Skylab, and the itinerant Space Shuttle, the Salyut and MIR stations paved the way for true long-term habitation and experimentation in microgravity at a very fair price. No one ever accused the Soviet Union of collapsing under the weight of its space program! The Salyut and MIR stations grew at a steady sustainable pace for over 30 years, most of the time, crewed and operating. The sheer economy of the project was an inspiration for western commercial efforts of the same size and complexity.

Examples of Experiments aboard MIR with Commercial Relevance -1986-1994

Experimental Discipline	Examples
Protein Crystal Growth	Catalase, gene engineered human growth hormone, luciferase, neuramanidase
Semiconductor, superconductor Crystal Growth	Cadmium selenide, gallium arsenide, cadmium telluride, zinc oxide and silicon - in the manufacture of semiconductors and high temperature superconductors. Germanium, yttrium-barium-copper oxide silver/germanium, antimony, lead-silver chlorides, aluminium/ nickel, gallium, gallium/antimony and aluminium/copper/iron. germanium, cadmium sulphide.
Bio-Processing	Human growth hormone, interferon, insulin, antibiotics, polyacrylamide, genetically engineered cells.
Space Agriculture	Wheat, ginseng, arabidopsis, spiderwort, saffron, stevii, flax, lettuce, peas, radishes, potato tubers, crustaceans, mollusks, fish, japanese quail, tree frogs, fruit flies.

The Shuttle and the MIR: Renewed International Cooperation

In 1995, the greatest achievements of the superpowers on Earth combined again. In June of that year, the Space Shuttle, Atlantis, docked with the MIR space station for the first joint mission between the Russians and the U.S. since Apollo/Soyuz in 1975. The die was cast for at least the next two decades. The arrangement was dicey at first. The Americans felt apprehensive as guests aboard the MIR. Equipment breakdowns, coolant fluid leaks, and the collision with the Progress supply vessel, are now legendary in space history. The project was an ugly success story, replete with humbling personal failures and the revelation of how decadent the Soviet Union had become.

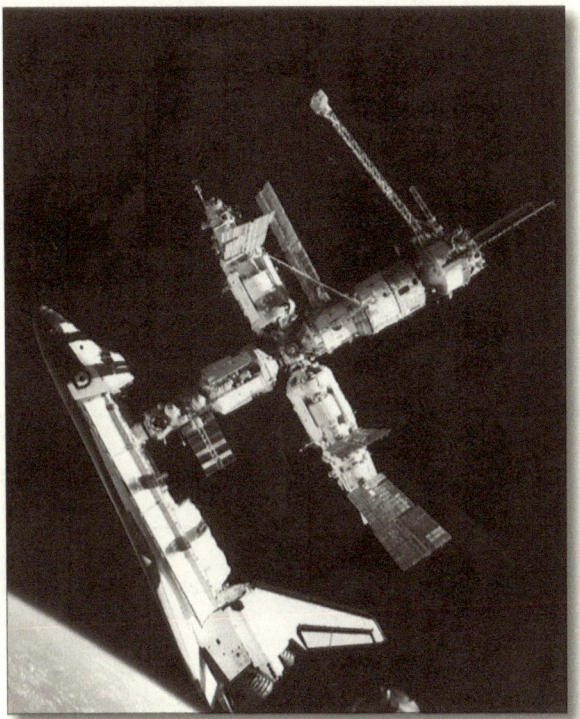

The Shuttle/MIR complex in 1995.
(Image courtesy of NASA)

Though the microgravity science conducted on MIR did not create any breakthroughs, certain experiments were successful. Wheat and mustard plants were grown, but did not reproduce perfectly in microgravity. Colloid crystal research had interesting results for use in optical displays, and was continued on the International Space Station thereafter. As long as astronauts exercise 2 ½ hours a day, muscle atrophy leveled off in about 4 months. Radiation studies confirmed that people are relatively safe in low Earth orbit. No long term illnesses have occurred that were attributable to space exposure. Collisions from large meteors never seem to happen. Only a small micro-meteoroidal "rain" persists that harmlessly peppers the outside of the station.

The summation of data between the two space programs tended to confirm each other expectations. The most significant was what was learned about Space Adaptation Syndrome. After some 35 years, certain conclusions could be drawn, all of them proving that short residence times in space are best. Astronauts must exercise regularly through this period. Six months in space is about the maximum. This limit is certainly adequate for commercial astronauts that are more concerned in saving timing than setting records.

Space Adaptation Effects in Humans

Symptom	Effects	Treatment
Vertigo, Motion Sickness	60%-80% of occupants	Anti nausea drugs.
Muscle Atrophy	Weakening of muscles in lower and upper body	Levels off in 4 months with exercise 2 1/2 hours per day.
Bone Loss	At least 1-2% per month. Does not level off.	No know effective treatments. Returns to normal after many months on Earth.
Hypercalcemia	High calcium in blood. Renal (kidney) stones could result.	No know effective treatments.
Exposure to solar, cosmic radiation.	Optic nerve impulses. No associated cases to illness later on.	No known completly effective barriers.
Blood Flow	Disperses to all parts of the body, making head feel puffy. Legs get colder if low temps are encountered.	N/A. Returns to normal on Earth.
Red blood cell loss		Returns to normal on Earth
Spine Length Increases	2-3 inches within days of entering microgravity	N/A. Returns to normal on Earth

Immune System Depression	Due to isolation and stress.	No know effective treatments.
Dehydration	Due to the body's new distribution of fluids in microgravity.	Increase fluid intake.
Head aches, congestion, insomnia	Puffyiness in face, degraded sense of taste and smell.	Drugs treatments on board
Space Adaptation Syndrome	Anxiety, depression, withdrawal, feeling of unease	Active work schedule. Regular exercise.
Weight loss	From loss of fluids, loss of appetite, muscle loss	On –board countermeasures of diet and exercise

Studies of plant growth were extensive on MIR/Shuttle missions. These experiments have a direct bearing on how space station occupants can cheaply and efficiently generate their own oxygen. Plants also have a pleasing aesthetic effect and can warm up the usually "spartan" spacecraft décor. Food supplies can be supplemented with gardens of natural spices and teas. Economical luxuries such as live plants will certainly be welcome on commercial vessels. Over many years of research, some crops were grown from seed to seed. Natural ethylene gas is created and needs to be ventilated to allow plants to flower. Irrigation water has to be pure and well contained at the root level.

Biological studies were not very promising at first. Animals have difficulty feeding in the bewildering microgravity environment and can easily panic. In reproduction experiments, birth defects were discovered in frogs and Japanese quails. No one knows if this was due to radiation or microgravity. Insects function well enough though, and bees make honeycomb nests, and spiders weave webs. It is not easy to care for animals in space and much work needs to be done to perfect microgravity biospheres for livestock. It seems that some gravity is needed here, and a rotating space station, with centrifugal force acting as gravity, should improve things for our animal friends.

MIR's Aftermath:
The Misfortunes of MirCorp and Mini Station 1

Only a few years after MIR was launched the Soviet Union dissolved. The $17 billion/year Russian space program became cash strapped. MIR needed to go. By the early 2000s, Netherlands entrepreneur, Jeffrey Manber, President of MirCorp, came up with the idea of commercializing MIR. The MIR would be a commercial habitat for scientists, tourist and filmmakers. It would also help fund a second $100 million dollar Mini Station 1 built by the Russian contractor, RSC Energia. Manber's MirCorp, a private company investing in extending MIR's life, even signed an agreement with the Russian Aviation and Space Agency, Roscosmos, authorizing a feasibility study for the Mini Station 1 and the use of the Soyuz spacecraft to supply it.

Many technological wonders were being dreamed up by Mircorp to advertise the station. Guests would spend about three weeks in microgravity doing all sorts of commercial activities, from alloy manufacture, protein crystal growth, and antibiotics research, to filmmaking, and even a Survivor reality show. Tourism was the main focus at first, and Dennis Tito, the first space tourist, was signed up to visit the station. Pop music star, Lance Bass was next in line and was training in Star City when the funding dried up.

After several promising rounds of financing by Mircorp, the Russian Space Agency decided to decommission the ailing MIR due to safety concerns and conflicts with ISS development. Energia also could not provide enough Soyuz vessels to supply both stations. There was a short time in which a leasing agreement was established, but this angered NASA, who believed that continuing MIR would distract Roscosmos from ISS development. This marked the end MIR. The Dennis Tito's voyage was taken over by the Space Adventures, Ltd. who arranged his trip to the ISS. The Mircorp initiative may have been too little too late. Russia has felt the burden of costly civil space stations long enough, and the U.S. would soon learn her lesson with the Columbia disaster in 2003.

The MIR saga shows us that the Space Race was indeed the silver lining of the Cold War. In retrospect, space has been a unifying place, even between rival nations with opposing ideologies. Ironically, the final battle was a capitalist struggle for credibility, with the Russian space station in the balance.

ISS Commercialism Today

The first four expeditions to the International Space Station were conducted in the business-as-usual manner. But with the George W. Bush presidency, an old theme began to emerge, one that is both heartening and challenging. NASA was called to an old task. It was time to fly into deep space again. No longer can it retreat into the nostalgic era of Apollo, or pretend to be a bastion of education for new generation of astronauts. NASA engineers were forced to face the future and become what they once were: explorers.

The ISS is still open to commercial and educational payloads as a "National Laboratory." In mid-2007 a NASA report was issued stating that the ISS must be used for its original science purposes 15% of the time. This was at the insistence of Senator Kay Bailey Hutchinson, of Texas. Though there is an educational budget at NASA that can cover those payloads, the commercial interests will have to pay their proportion of launch fees and astronaut time. This is a bit of a "catch-22" with Shuttle launch costs as expensive as gold ($10,000/lb) with astronaut time fees added to it!

The Vision for Exploration, however, provided exactly what NASA was complaining it lacked for so many years: a clear directive. The space station science program isn't sexy enough for taxpayers. Even when the New Vision was plainly justified by Congress, the scientists who counted on space research opportunities were appalled. Most of the experiments planned for the ISS were being differed of cancelled. But this was the best medicine for NASA since Kennedy's famous Moon speech. It delineates what NASA can do, and what it can do without. The agency should not be landlords for space station chemists. This job is best accomplished through private industry so that space colonization it can be sustained.

Scientific research aboard the ISS will not go away. The European Space Agency (ESA) and Japanese Aerospace and Development Agency (JAXA) and Roscosmos will continue to provide research opportunities for civil and commercial needs. Their respective leaders also believe that science can aid in bringing about a better society. The ESA and JAXA will assist Roscosmos in re-supplying their own modules using the new automated cargo vessels, Jules Verne (ATV) and HTV, respectively.

The Russian segment of the ISS has been quiet lately, but Roscosmos doesn't plan to keep it that way. In mid-2006, RSC Energia, the main contractor for Roscosmos and the builder of the ISS, MIR and Salute space stations, reaffirmed that it will continue to build up the ISS if funding is provided. Sometime in 2008, the new Russian segment called the Multi-Purpose Laboratory Module (MLM) will be launched. It is intended to be a fully equipped commercial facility that will help pay for itself through industrial and scientific research. Foreign scientists will be welcome to use it for consumer product development. It will also be a staging platform for future Moon missions, and the support the development of a reusable space plane that will bring up four paying customers as one time.

The other ISS partners have been influenced by America's directive to pursue exploration science. Russian Mars and Moon missions are back on the drawing boards. NASA also has not entirely given up on commercial research either, and continues multi-purpose experiments on the ISS in pharmaceuticals, bioprocessing, materials research, colloids, agriculture and others that have dual purposes and commercial relevance. Spin offs still trickle down from the ISS, and new ideas keep tumbling in from all directions. The big missing piece is in a separate commercial habitat that can help with the overflow, and cut through the NASA bureaucracy to meet tight schedules.

Commercially Relevant Microgravity Experiments on the ISS (2001-2006)

Expedition	Experiments:
1	Protein crystal growth
2	Protein crystal growth, bioprocessing, agriculture
3	Protein Crystal Growth, exposure of materials to space, video, colloids, biotechnology - cell growth
4	Protein crystal growth, agriculture – photosynthesis, exposure of materials to space, colloids bioprocessing – antibiotics, cell growth
5	Zeolite growth, semiconductor crystal growth, biotechnology – liver cell research, agriculture protein crystal growth, materials exposure to space, microencapsulation
6	Zeolite growth, agriculture, protein crystal growth, materials exposure of materials to space, colloidal physics
7	Agriculture, protein crystal growth, metal/alloy research, soldering in microgravity, colloid physics metallurgy physics, biotechnology - cell growth
January 2004	**New Vision For Space Exploration Announced**
8	Biotechnology - cell growth, fluid merging viscosity measurement, binary colloid physics, space soldering experiment, exposure of materials to space, agriculture, protein crystal growth
9	Binary colloid physics, foamed metallic glass, soldering in microgravity, agriculture, protein crystal growth
10	Binary colloid physics, biotechnology - cell growth, exposure of materials to space, agriculture, protein crystal growth
11	Motion sickness research, exposure of materials to space. agriculture, protein crystal growth

12	Binary colloid physics, biotechnology - cell growth , exposure of materials to space.
13	Binary colloid physics, biotechnology - cell growth, exposure of materials to space, agriculture pore formation is molten metal.
14	Agriculture, exposure of materials to space, skin care studies (ESA), small satellite tests

Experiments of China's Manned Space Program

Arriving late in the manned space flight scene, the China National Space Administration (CNSA) has no illusions. They went right into the business of trying to benefit the nation with consumer goods. Instead of believing that manned, on-orbit facilities are merely military observation platforms, astronomical observatories or communications stations, they concentrate on the benefits microgravity can provide the Chinese economy.

The Shenzhou 5 manned space station launched in late 2002 was not just a test flight. Scientists at the China National Space Administration (CNSA) were more interested in the microgravity science aspect. The CNSA flights are very similar to Russian Soyuz missions. The hardware and architecture allows for a slow evolutionary path of development with a minimum of manned launches, and therefore, risks.

The CNSA began microgravity research with unmanned retrievable capsules. These would be launched into orbit, circle the Earth for a few weeks, and come back down. The Soviet Union also conducted retrievable capsule missions two decades ago as an adjunct to space station research.

One interesting discovery has been in crop breeding. Since 1987, CNSA scientists have sent more than seventy kinds of crop species on eight space missions. They successfully cultivated a series of high-yield and high-quality agricultural products, including rice, wheat, tomato, green pepper, sesame and potatoes. Because of the combined effects of radiation and microgravity, exposed crops seeds have yielded a 10%-25% increase in weight of peppers, tomatoes and rice.

At present CNSA is the leader in making retrievable unmanned space stations a valid research tool. Recent retrievable spacecraft also carried aboard a multi-chamber space crystallization furnace, protein crystal growth equipment, a cell bioreactor, and electrophoresis device for chemical isolation and refinement. No commercial space company in the west can send a payload into orbit, keep it steady of a month or so, and then bring it back down intact. Yet today's protein crystal growth researchers need just this capability to continue their promising medical experiments.

The manned Shendzhou 5 space station mission in late 2002 carried aboard one taikonaut and a battery of microgravity tests for medical research. While only in space a short time, tests were conducted on fluid physics, electrophoresis, and the fusing of animal and plant cells to design antibiotics. The CNSA's focus on agriculture is in contrast with the American focus on biotech and the Russian inclination towards alloy research. Each has something to offer the other in trade through their diverse in cultural perspectives. The successor spacecraft, Shendzhou 6, reached orbit with two taikonauts aboard in 2005, with a new flight scheduled for 2008.

The sheer cost of space travel alone necessitates international cooperation, even between incompatible cultures. As the ISS drifts over the Middle East, it symbolizes how western cultures, with no precedents for peace, chose to work progressively for a positive future, and not back into their ancient tribal pasts. In 2006, an Iranian-American woman entrepreneur joined the cadre of German, American, Russian astronauts in a display of unity and reconciliation. It shows our struggles and hatreds are not necessarily inherited down the generations.

A commercial space station, apolitical and international, can help bridge the cultural chasms, east, mid-east, and west through mutual interests of new products and services. It establishes a partnership that costs more to disrupt than to maintain. The technological problems of Earth can be solved technologically. The new physics of microgravity, developed through a marketing perspective, can help turn us towards more optimistic pursuits rather than down the dismal path that some see ahead.

3

THE NEW SPACE CONQUEST: PIONEERING MANNED COMMERCIAL STATIONS

THE DREAMS OF MANNED SPACE stations go back as far as the late 19th century, when American writer, Edward Everett Hale, published his famous science fiction story, "The Brick Moon". This story, printed in the Atlantic Monthly in 1870, was told in the classic Victorian style of H.G Wells and Jules Verne. It described a manned satellite, made of bricks, that was flung into space by rolling down it a hill and into two huge, spinning flywheels. Their combined power vaulted the Brick Moon into an orbit over the north and south poles. The "station" was large enough to house thirty-seven people, but its purpose was navigation, principally for the fishing industry.

Mariners of the 17th and 18th centuries were beset by the difficulties in getting a correct fix on longitude. Latitude could be derived from the sun's position with a sextant, but longitude relied on compasses that were inaccurate. Crews starved as ships were lost trying to find their way through routine trade routes. In addition to the human tragedies, insurance companies had to shoulder the burden of material losses as well. It wasn't until the development of the chronometer in the mid-1700s that sailors had real control over their locations at sea. This and other navigational mechanisms were expensive though, and the impoverished fishermen of the world could not afford them, even in 1870.

An artificial satellite, orbiting around the Earth from pole to pole would give any navigator a constant fix on longitude. This was not only the first idea for commercial space stations, but one of the first for navigational and communication satellites well.

The Brick Moon was designed to fly unmanned. It was 200 feet in diameter and hollow inside. It was to be whitewashed so that people could see it on Earth. The cost of the project was $162,000, privately raised and organized by a generous philanthropic businessman. Though the satellites would aid in commercial fishing, no profit taking was intended. The entrepreneur who organized the venture would gain a premium only in his heart.

> "For he believed, on his soul, that the success of this enterprise promised more for mankind than any enterprise which was ever likely to call for the devotion of his life."

Then "ground control" had a problem. While the Brick Moon was being constructed, it slid down he hill. With thirty-seven people onboard, the half finished moon was launched to an altitude of 5109 miles! With despair being the "chief of all sins", the inhabitants stoically made the best of it in space. In 1870, people still believed that Earth's air and rainfall existed that far up, so the colonists lived comfortably in the "tropical climate" of space. They communicated with Earth by jumping up and down in "Morse Code".

> "They were telegraphing to our world, in the hope of an observer. Long leaps and short leaps, -- the long and short of Morse's Telegraph Alphabet, -- were communicating ideas."

In this quaint yet progressive tale, space entreprenuerialism, navigation satellites, manned space stations, and cost-effective methods of launching and communicating with spacecraft all made their pop culture debut. Just as the first serious plans for Moon voyages came out of 19th Century writers, so did the idea of manned space stations.

The idea had a lot of traction. In 1903, a Russian high school teacher, Konstantin Tsiolkovsky considered space stations as a logical staging point before going on to other planets. In 1928, Herman Potocnik, (aka Hermann Noordung) a Croation military captain, created a detailed design on what a space station might look like. His devised a wheel shaped vessel that rotated slowly to produce centrifugal force so that travelers could enjoy artificial gravity along the inner skin on the stations rim. But the idea for space stations really hit stride after the world wars when aviation technology had accellerated enough to make space travel considerable.

Probably the most notable and graphical depiction of manned space stations before the 1960s was from the now legendary series of articles published in Colliers magazine in 1952. In these articles, NASA's agenda was set in stone to this very day. Wernher von Braun, and science writer, Wiley Lee were the authors, and Chesley Bonestell, the famous renderer of the Chrysler Building's facades, was the illustrator. It was a bombshell on the world of popular science. Many a space engineer was born that year. Based on the Colliers work, these articles, wrtten by Wiley Lee (author), Joseph Kaplan (physists), Wernher von Braun (Guided Missile Development Group Technical Director), Heinz Haber (astronomer), Oscar Schacter (United Nations lawyer), and Fred Whipple (astronomer), and were expanded and bound in a gloriously illustrated book, *Across the Space Frontier* which showed a level of detail in space station hardware unprecedented before then in the popular media.

Borrowing from Herman Potocnik's concept, Von Braun's space station was the wheel-shaped variety, the published details of which defied description. Every system and subsystem is accounted for, including solar energy for power, rotation for artificial gravity, heat, cooling and water recovery plants, and transportation using a shuttle-like winged spacecraft. Arthur C. Clark used this design in his novel, *2001: A Space Odyssey,* which was later depicted in Stanley Kubrick's breakthrough classic film. Though some said it had a lame script and confusing story line, the backdrop was all Von Braun, Bonestell, and Clark.

But it wasn't the first film depicting a rotating space station. Hollywood producer, George Pal created what could be considered the predecessor of the 1969 film.

The Conquest of Space, made in 1955, showed a Potocnik –style space station in full operation. Its purpose was as a staging area for a trip to the Moon. In the movie, the primary goal was superseded in favor of a trip to Mars. This was quite a surprise to everyone, and chaos ensued, including the homicidal madness of the station's military commander, and a nail-biting launch from Mars back to Earth.

Though less than a masterpiece, the film does document the ideas prevalent of the day: a strict military command structure, international crews, and the belief that resources could be mined in space to help solve the overpopulation problem back on Earth.

After 41 years from the launch of Sputnik, a permanent manned space station project finally began in 1998. The dream of the early twentieth century rocket pioneers unfolded in the 2000s, one hundred years after its first serious concepts were drawn up. The original plan was to make it an all-American station, but the cost was too great for one nation. Early in the planning stages, the space station was about to be cancelled by the American Congress. Inspired reasoning prevailed. A changing world economy opened new ways of thinking about space markets.

After the fall of the Soviet Union, the U.S was greatly concerned about the Russian government selling liquid rocket hardware to foreign countries. In 1993, it dawned on Yuri Koptev, leader of the fledgling Russian Space Agency, that if NASA cooperated on a joint space station, both countries could continue with their space plans without excessive economic strain. It would also give the Russians an incentive not to sell rocket technology on the international market. This also gave a publicity boost to the "New World Order" that had not yet brought about a tangible benefit to the Russian economy.

NASA Administrator, Dan Goldin liked the idea and proceeded to sell it to the Clinton Administration. Instead of canceling the space station, NASA now had a partner in construction and transportation. The ISS was born out of mutual necessity.

A joint American-Russian Shuttle/MIR mission was a practical starting point. The first stage was Phase One: finishing up the Russian's space station MIR program. Phase Two was to build an International Space Station (ISS) from scratch, but the Russian Space Agency had a hard time finding the cash to get the first segment in orbit. After many frustrating delays, the Functional Cargo Block, Zarya, funded by the U.S., was deployed in 1998. The American bailout had finally bared fruit.

After things were up and running, the tourism controversy began. The first self- funded astronaut, Dennis Tito, paid over $20 million to spend a week on the ISS. The Americans objected, saying that the station was not finished and that an amateur could endanger the crews' safety. The Russians insisted, and to this day, tourists hitch a ride on Soyuz spacecraft to visit the station. This was huge step in the commercialization of manned, on-orbit activities. Without it, there would be no objective opinions of how great space flight is, and its credibility would be continually challenged. The Russians bailed out the commercial sector in historic fashion amid the objections of an ostensibly "pro-commercial" NASA.

Launch support has always been an attractive capability of the Russian space agency. The Soyuz, a personnel and cargo vehicle, is reliable and affordable. Without it, there would be no ISS. This was never more evident than when the Space Shuttle, Columbia went down in February of 2003. If skeptics ever doubted the need for international cooperation is space, their objections were quashed by this terrible tragedy. It wasn't until mid-2006 before the next Shuttle assisted phase of space station construction could took place. In the meantime, the Soyuz kept the ISS alive and inhabited by crews.

The ISS is not like the wheel-shaped fantasy depicted in films, but NASA was back on the course Wernher von Braun had proposed for America long ago. But as it turns out, you may not need or want artificial gravity in space stations where industrial research is being performed. It most cases, it defeats the purpose of being in space at all.

The High-Cost-High-Return Model

Though artificial gravity is desirable for creature comforts, the money is made in the weightless space environment. The first objective of the commercial space station is to utilize microgravity for chemistry and physics applications. The agenda is to build a stable platform that will be available for years. Guests will only stay for a short time, with experimental results being continuously off loaded, and with robots playing a large role in the day-to-day processes. The customer base will be international and multidisciplinary in the arts and sciences.

But how do you make such a mind-boggling investment pay off? First of all, there is already an extensive stream of paying customers for suborbital flights. The Virgin Galactic SpaceShipTwo, to be flown in 2008 on short pleasure flights, has a list of over 100 passengers already. Assuming you have enough customers, how does anyone profit from manned space activities? The operation costs are still going to be too high, right?

Not necessarily. One approach is to cover the transportation expenses for the customer. Costs for the Soyuz human rated rockets are $40 million per flight (according to the FAA), with some vehicles in development that will make launches somewhat cheaper. You don't want to pass on the high prices to the customers if you can avoid it. All you want the customers to pay is a rental fee for the compartments used on the station if possible.

How? By providing the largest amount of rental space you can, and packing it full of as many payloads as you can. After a few years, the station pays for the transportation and deployment costs by leasing its compartments. This is the same model used for satellites today. All commercial communication satellites are built, launched, and leased to as many customers as possible, as long as possible.

For example, the average university sized payload might be an alloy-blending furnace about two cubic feet in size. It has a six-month minimum residence time. Two cubic feet is leased for $120 per day or ($60 per cubic foot/day). If the apparatus is in space for 183 days, it will cost the customer only $21,960 for six months. By today's standards for payloads, these fees are peanuts by comparison.

Now let's see how much a hotel room would cost tourists. They can float around anywhere in the unrestricted areas, but they have to live in the rooms that they rent. A cabin 25 x 25 x 10 for two weeks is $5,250,000. Compare this with the current costs of over $20 million for one week! But $5 million is still pretty high. Could the fees be made even lower?

Governments throughout the world give out grants for scientific research. If the ISS has been decommissioned, and American, European and Japanese space agencies stay out of the space station business, where is space research going to be performed? Why own when you can rent? For a fraction of what world governments would pay to build, operate, and supply a new International Space Station, they can help industry pay for microgravity research by renting space.

The difference is dramatic. If the international partners spend $100 billion on the ISS over a period of eighteen years, the station would cost approximately $5.5 billion per year. Consider a commercial space habitat with 70,000 cubic feet of habitable volume. Even if the world governments rented the entire station, it would cost them only $1.5 billion per year. (70,000 cubic feet x $60 per day). By sharing the costs with industry, it would be even lower. Since the commercial space station is busy creating products, the taxpayer would be paid back in products and services. Taxes are then levied on commercial sales and the government is paid back. The station becomes more of a tax base than a tax expenditure.

The European Space Agency has already begun this process for business development on the International Space Station. Financial assistance is offered for entrepreneurs and veterans alike, complete with on-line application forms. Instead of grants, the governments lend money or help users with differed payments, partial payments, and even complete, start-to-finish development contracts if the payload benefits society as a whole. Disease control, energy research, etc. are examples of philanthropic research goals that may garnish full political and financial support.

As stated before, the key to profits in space is leasing out the largest possible volume to customers. This means you have to put up a giant space station in a very short period of time. The station must be tough enough to last at least ten years without a mishap. It has to be kept very steady to protect experiments in progress. One way to do this is to outfit a simple rocket fuselage that can be launched in one piece, and started up right away. Fortunately, we did it years ago. It was called Skylab.

The Skylab program began 1960s alongside Apollo. It was in the spirit of the Von Braun agenda to build a permanent space station and segway to the planets. Converted from an empty Saturn IV-B second stage booster, the hollow shell was outfitted with a turnkey space workshop. It was built and tested on the ground and launched in one piece, ready to go. A subsequent launch accident (see Chapter 2) delayed the mission after an entire solar power panel was torn off. But the windmill like solar arrays, which powered a solar telescope, unfolded perfectly, and the station continued as planned.

Skylab is a model for commercial space stations. It was built with surplus equipment, deployed with one launch, and was supplied by tried and true vehicles. The last available Saturn V Moon rocket was used to launch the station, and some of the remaining Saturn 1Bs rockets supplied it with three crews. If NASA had not discontinued the Saturn booster line, Skylab would have been inhabitable for several years in a fashion similar to the Russian Salyuts.

The use of fuselages for space stations goes back many years. NASA Directors George E. Mueller and Wernher von Braun championed the idea in the mid sixties. They argued that Skylab could be launched empty and furnished with all its equipment while in orbit. Although much more difficult to achieve, a successful on-orbit outfitting would really advance human proficiency in space. The idea was hotly debated, but it became clear that the program might be cancelled if it were too complicated. By 1969, Skylab was redesigned from a space-built station to one that was outfitted on the ground.

The orbital outfitting concept is alive and well today. With the cancellation of the Space Shuttle in 2010, the big orange fuel tanks used by the shuttles will be discontinued. However, the same companies who built them will be building the new Aries 5 tanks. These are much larger than the Shuttle tanks and can also be converted into space habitats.

One of the most persistent lobbyists of fuel tank space stations today is Gene Meyers, President of Space Island Group, Inc. He and his engineers have developed methods of building, launching, supporting, and financing space stations for several years. His principle goals are to use them for tourism, research, manufacturing, sports and entertainment. His plan is to make billions, while absorbing the costs of transportation.

Space Island Group: Skylab on Steroids

Gene Meyers, a retired engineer living in West Covina, California, founded Space Island Group Corporation in 1992. His company is devoted to developing habitats in low Earth orbit. Back in the early 1980s, Meyers read about an idea to convert an empty Shuttle fuel tank into a space station. After a few years of developing various designs, Meyers quit his position at TRW, and started his own company to work on the idea full time.

Space Island Group Corporation was born. Meyers' fundamental concept is to use rocket fuel tanks as orbital "real estate". They are to be the building blocks for commercial orbital complexes consisting of stations, launch vehicles, telerobots, solar satellites, and fuel depots.

Meyers' reasoning is simple: pay for the cost of building the space stations by leasing the enormous habitable volume to customers. The empty space inside of a fuel tank is huge, much larger than the ISS. If this jumbo Skylab were packed full of industrial products and services, it would pay for itself in a few years. The attractive aspect is that his lease rates are very reasonable: $25 per cubic foot per day, transportation included.

Through the last 30 years, the most difficult hurdle for space commercial entrepreneurs has been launch costs. The simple matter of getting any payload from the surface of the Earth to an orbit 200 miles up is very daunting. The average booster today is about $75 million dollars per launch, and this can go up as high as $180 million. However, these prices are only for communication satellites that go to very high orbits of 10,000 to 23,500 miles.

For manned flights, the cost is even higher, and for a much lower altitude. Payloads have to include life support and abort systems, making the entire vessel cost at least $40 million just to bring three astronauts 200 miles up. The Space Shuttle is $400 million for seven astronauts. It is now over $20 million dollars each to launch a commercial "guest" astronaut to the International Space Station. There is no way to make money if the initial start-up launch costs are this high.

Meyers believes he can change all that by simply boosting up large volumes of living space at one time. The more real estate you have in space, the more you can charge per day. Even at $100 million per flight, the costs could be made up by the high amount of living space leased. Space Island Group also requires a 10% royalty fee on every product or process invented on the space stations. Gene Meyer's "zero cost to orbit" business model is used today for communication satellites: as long as the payload generates high amounts of revenue, it does not matter how much the launch costs are as long as they are recoverable within the operating lifespan of the spacecraft. This is the high-cost-high-return business model discussed earlier.

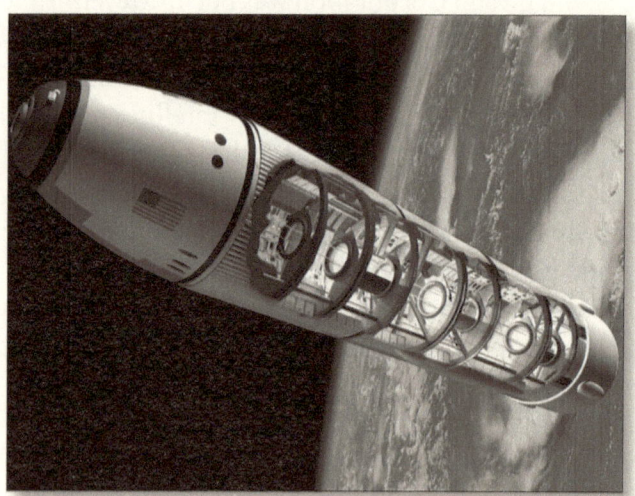

Skylab on Steroids. Each Space Island Group module is a turnkey space station launched fully assembled. *(Image courtesy of Space Island Group)*

"People think that the NASA prices are the true prices", says Meyers, speaking about the current multi-billion dollar cost of building rockets. "NASA makes everything to be used only a few times, and then they are on to something else. If you mass-produce these components, the price drops dramatically. Most of these components are based on military specs for really customized uses."

Shuttles and their fuel tanks will no longer be built, but the new NASA designs for Ares vehicles use the same technology and will only help Space Island Group design its latest concept, the Dual Launch Vehicle. The large fuel tank of the Aries booster will give the stations even more habitable volume than the Shuttle tank. It can be configured for 150,000 cubic feet for subsystems, crew, and cargo or laboratory space. Hardware can be reconfigured in different ways, and can be mass-produced. Space Island Group believes they can issue their own launch vehicles. These proposed rockets are simple and expendable. They would cost $5-$6 billion to deploy, compared to Space Shuttle's initial $15 billion price tag.

The Dual Launch Vehicle on way to orbit. Engines are mounted on the bottoms of the fuel tanks. The white cargo vessel (right) carries supplies, crews, and return vehicles. *(Image courtesy of Space Island Group)*

"In our business plan," says Meyers, "we assume that launch cost for our vehicle, which is essentially two external tanks launched side by side; one full of fuel, one full of cargo. The external tank [that] we will use will be a stretched version of the one NASA uses. We can lease it out for $25 per cubic foot per day [and] get all our launch costs back in 9 months."

With these new concepts, space travel doesn't have to be as expensive as everyone assumes it is. With mass production, low lease rates, proven hardware, and experienced aerospace personnel to help with licensing and technical support, Meyers feels the road is already paved to begin developing.

And who will actually build these space stations? The same people who build shuttle fuel tanks.

"Lockheed Martin can make these now," says Meyers. "One launch of our [Dual Launch] vehicle will put three times as much interior volume up there [in orbit] as the entire International Space Station."

The plans Space Island Group has for the near future are not conservative. It is more like Author C. Clarke's *2001: A Space Odyssey* than any other scenario. But there still is one serious issue yet to be resolved: how to make quick round trips to the space stations. Space Island Group rockets are all one shot, expendable launch vehicles. Once astronauts get up into orbit, how do they get back down?

Back in the early 1990s, a unique development took place that created a fresh interest in commercial space flight. Engineers, then at the McDonnell Douglas Corporation, now a part of Boeing, started tinkering with an experimental vessel called the Delta Clipper. This remarkable cone-shaped vehicle flew up into the sky and came down again much like the rockets shown in old 1950s science fiction serials. This "back to the future" concept enthralled the public and the space flight community alike.

Everything changed overnight. By 1995, dreams of fast intercontinental travel, cheap tourist trips into space, and same day package deliveries, danced in the heads of entrepreneurs and communication moguls alike. Even Bill Gates wanted to build an "Internet in the Sky", called Teledesic, using 840 little satellites launched by small, reusable launch vehicles. The small sat market failed to materialize, but in the meantime, a slew of reusable spacecraft designs had emerged. Thus began the revolution we see today in suborbital spacecraft like the one being built by Virgin Galactic.

Led by Dr. William Gaubatz, the Delta Clipper could move laterally in space, slowly descending and landing on four legs. It was not capable of real space flight at that time and was only an experiment. Space Island Group has now revived the idea. It fact, the company needs this vessel to make the fuel tank stations work without the Soyuz or Shuttle to supply them

Funding for the Delta Clipper ended soon after its first test flights. But the idea for an orbital passenger ship never died out completely, and to this day, the dream is alive and well with companies; such as, Pioneer Rocketplane/Kistler, SpaceX and Dreamchaser who wish to bring down the cost of space launches to an affordable level.

Space Island Group didn't forget about the Delta Clipper either, and wants use the spacecraft as a return vehicle only. Gene Meyer envisions the small Clipper spacecraft being stacked on the top of fuel tanks and launched as part of the Dual Launch Vehicle. The tank

itself will be a fully assembled space station with solar panels and life-support systems. Mounted on the on top of the lab will be a Delta Clipper that can bring astronauts down from the space stations. The Delta Clipper can even land in a gentle fashion so as not to damage the delicate products made in space. It also provides orbital maneuvering capability for the crew.

Space Island Group's Founder, Gene Meyers, and his model of the Dual Launch Vehicle.

These stacks of vehicles are designed to be launched and landed on a regular basis, and the boosters' empty fuel tanks can be used for future stations. In just one launch, two space stations and a landing craft can be sent into orbit, ready for use. By 2028, low Earth orbit could be festooned with orbiting fuel tanks, outfitted for use as chemistry labs, movie studios, or hotels.

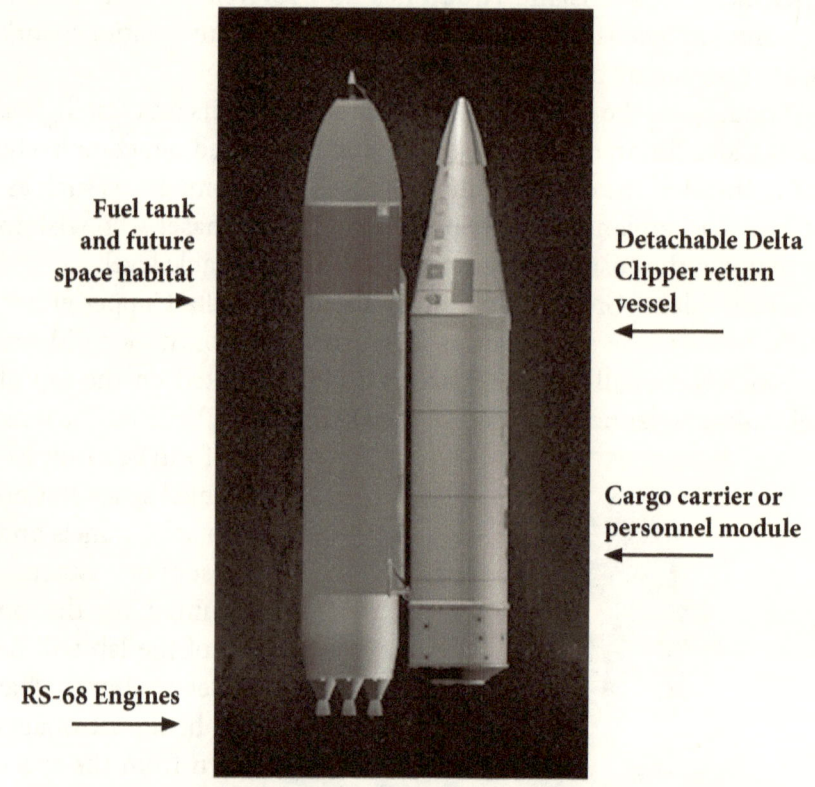

Fuel tank and future space habitat →

← Detachable Delta Clipper return vessel

← Cargo carrier or personnel module

RS-68 Engines →

The new Space Island Group stack shown here without solid rocket boosters *(Images Courtesy of Space Island Group)*

Meyers already sees a way of making the trips even more lucrative. Why not send more than one Delta Clipper up at a time?

"The cargo vessel can be expanded so that we can take up to three Delta Clippers with each flight," says Meyers. "You could actually put two of these inside [the cargo carrier] and then a third on top."

The Delta Clippers would greatly enable astronauts to get around in space and dock with other fuel tanks stations. As long as an access hatch to the empty rocket is available, a habitable volume can be pressurized. In the stack above, a hatch would be built into the side of the fuel tank where astronauts would be able to get in and refit it for use. Eventually many habitats could be attached together in a cluster for microgravity research.

These high cost/high return economics could help cash starved universities achieve regular unmanned access to space. By 2012, a small, microwave oven-sized payload could be launched, operated for six months, and retrieved for under $30,000, instead of the millions of dollars it takes in 2007. An 18 cubic foot rack used for automated protein crystal growth experiments would cost only $165,000 per year. A small movie studio 25 x 25 x 10 would cost under $60 million per year, instead of the current price of $20 million per week!

Another development issue to be resolved is the infamous foam insulation on the outside of the fuel tanks. A piece of orange foam broke off and caused a puncture in the Space Shuttle Columbia's wing during its fateful launch in 2003. The damage caused the spacecraft to burn up upon reentry killing seven astronauts and grounding the fleet for two and half years. On the Space Island Group's Dual Launch Vehicle, the insulation might hit the adjacent payload upon launch. It will also break off while in orbit, contaminating space with particles that could eventually turn into tiny bullets as they pick up speed and race around the earth. Such debris could severely damage or contaminate weather and imaging satellites as well as the International Space Station.

Space Island Group's solution is to use a double hull. The booster will be made the same way as the Ares is, but a new kind of foam will be used: Aerogel. Also known as "frozen smoke", this lightweight plastic is an excellent insulation material for spacecraft. A layer of Aerogel would be placed in between the full tank and the outer metal hull to protect against meteor damage. Similar to Skylab's meteorite shield that was torn off on launch, the double-hull design will also keep astronauts warm and cool by providing a jacket of protection from the sun and deep space.

Space entrepreneurs will also need an Orbital Maneuvering Vehicle. These are flying robots that are "joy stick" driven from inside the space stations or from Earth to perform tasks in space. They will minimize the need for space walks by astronauts. The small robots are based on old NASA concepts and have radar directional controls, camera view ports, and large mechanical grappling arms that are versatile, reliable and dexterous. They would be needed to retrieve satellites, build solar panels, and assist with all the dangerous chores that commercial astronauts face. According to Meyers, the development costs for Orbital Maneuvering Vehicles are projected at $200 million, eventually costing $10 million per robot.

The use of telerobots whenever possible *must* be a tenet of the new commercial space age. Without these robots and other telerobotic devices, space habitats will be challenging, dangerous places. Paying for astronaut training and services, as well as spacesuits, would change the high-cost-high-return model considerably, because the salaries of the astronauts are at a premium. Should the attendant astronaut die or get injured, legal action could shut down a space station entirely. Though astronauts must be around to fix the broken equipment, the telerobots must be doing most of the work to make commercial space economical.

New launch facilities will be needed for the Space Island Group venture, so Meyers wants to build a spaceport right near the Kennedy Space Center site that supports the Shuttle today. Much of this area is run down because actual Shuttle launch operations were moved to an adjacent sight. Space Island Group would refurbish the site for its own vehicles. Meyers is confident that full scale facilities, much like those already in use at the Kennedy Space Center, would cost $200-$300 million.

Because the business development phase is just getting started, Space Island Group is interested in other space companies to become customers. Meyers says he will gladly team up with other industries for launch support, space station interior design, upgrades, and return vehicle improvements. Rocketplane/Kistler K1 and SpaceX's Falcon 9 boosters could be used support the Dual Launch Vehicle stack by replacing the solid rocket boosters and return vehicle, respectively. Bigelow Aerospace was invited to use their Kevlar-like fabric to line the inside of the space stations to protect passengers from radiation.

""We're really not competitors with those guys," says Meyers. "When the SpaceX engines get reliable enough, we will replace the solid rocket boosters with Falcon engines. We will be their biggest customer, much bigger than NASA. Bigelow's [inflatable] material would make ideal linings for the interiors of the living quarters"

The ultimate goal is making space launches free so that they are never a burden to the customer. But there still will be the need for initial financing in any case. Space Island Group needs $10 billion to cover the initial costs.

Fortunately, Gene Meyers travels in the kinds of circles that can give him large amounts of seed money. His investor audiences consisted of a soft drink company to sponsor sports arenas in space, an aluminum company that would use the recycled materials from old satellites retrieved by space island robots, casino developers who would license Space Island Group themes for new Las Vegas hotels, NASA, where low gravity environments are needed to simulate Moon and Mars missions, and most lucrative of all, the military for energy production in space. While at the Pentagon, Meyers had an interesting conversation involving the future of warfare.

"When talking to the generals on what would draw American troops into combat in the next 20 years," says Meyers," all the causes were energy related."

Meyers' new objective to finance space stations is to create a solar power satellite system. The concept, first conceived of in the 1960s by Peter Glaser, is a large solar panel array that can collect limitless amounts of energy from the sun and send the power down to Earth in the form of microwaves. Solar Power Satellites, built by the Orbital Maneuvering Units, could provide energy for the emerging power grids of India or China. Rights to the power can even be pre-sold to provide the $10 billion in start up costs for the Space Island Group stations. (See Chapter 5).

High launch costs are the main reasons we do not have solar power satellites today. Space Island Group believes that if space stations could pay for themselves through industrial leasing, the launch costs for building the solar power satellite would be reduced by packing solar satellite building materials in with each cargo flight to the station.

Bigelow Aerospace and the Great Space Gamble

In July of 2006, Bigelow Aerospace deployed the first privately funded commercial space station. Orbiting at a 350-mile altitude, the 14.4 feet x 8.3 feet foot space station showcase is a prototype that will open opportunities for all types of investment, from advertising to microgravity science. This spacecraft, Genesis I, is an inflatable module made of a Kevlar-like fabric that is tougher than aluminum. Larger versions of this inflatable are in the manufacturing stages and will be manned by commercial astronauts.

Ballooning to thirty-eight cubic feet of interior volume, the station sports several external and internal cameras, one of which peers down to the Earth surface and snaps images. Solar panels are included for power and communications, along with experiments and commercial items used for advertising. Traveling at 17,000 mile per hour at a height of 350 miles, the Genesis 1 is designed to be an unmanned pathfinder mission for advertising and feasibility functions only. No astronauts will board this spacecraft and it will not return to Earth. It will burn up in the atmosphere after several years.

Actual Genesis 1 photograph taken aboard the orbiting spacecraft.
(Courtesy Bigelow Aerospace)

This remarkable economical feat was made possible with a Russian rocket known as the Dnepr. Serviced by the ISC Kosmotras company in the Ukraine, this launcher was a descendent of old SS-18 ICBM rockets left over from the Cold War. The Dnepr was launched from an underground missile silo in Dombarovsky, Russia. The vehicle costs about $9.5 million to launch, as opposed to the average vehicle's price tag of $75 million. The Genesis I was folded into the nose cone of the rocket and was inflated once it reached its altitude.

The President of Bigelow Aerospace, Robert T. Bigelow, is a graduate of Arizona State University where he got his BS degree in Business Administration. Owner of the Budget Suites Hotel chain, he founded Bigelow Aerospace with his own fortune in 1999. He, like Howard Hughes before him, is an aerospace engineer by nature, not by nurture.

His business strategy is simple: use lightweight inflatable spacecraft instead of the heavy aluminum cans NASA uses. This will lower the costs of launch and allow for the building of human habitats in space from low Earth orbit to the surface of Mars. If these inflatables prove as tough as aluminum after a reason amount of time in space, they could revolutionize how humans will inhabit space in the future.

Bigelow Aerospace is a small company with only about 120 employees. It has production locations in North Las Vegas, Texas, , and Washington, DC, and a corporate headquarters in the heart of Las Vegas. The company has its own Mission Control Center where technicians track the motion of the spacecraft, and make adjustments with desktop computers. As of 2006, $75 million has been invested in Bigelow Aerospace to gain affordable access to microgravity environments.

Following the Panamsat business model of 100% private funding, (Discussed in Chapter One: Satellites: The First Commercial Space Stations), Robert Bigelow has pulled off a major breakthrough. The launch of his Genesis I was an historic event that did not capture the interest of the press as the more photogenic SpaceShipOne suborbital flight did in 2004. The Genesis I, however, was even more significant. It has proven the technical credibility of accessible commercial platforms in space where experiments in microgravity, and all other activities, can be performed.

What makes Genesis I so lightweight, as therefore, so affordable, is the type of material it is made of. It is a composite of several layers of a Kevlar polymer laminated to create spacecraft hull six inches thick. This is the same material in the bullet-proof armor worn by soldiers. It can even be loaded with packets of water 2 1/2 inches deep to help block radiation from space.

"Our vessels are designed to be more of an inhibitor of radiation than the International Space Station is designed to be", says Bigelow.

In an August 24th, 2006 radio broadcast on Dr. David Livingston's, The Space Show (www.thespaceshow.com), Mr. Bigelow emphasized the amazing strength of the Kevlar material. Ironically, by using Kevlar inflatables, rather than aluminum "tin cans", spacecraft safety is actually enhanced rather than compromised. When meteorites hit Kevlar they shatter on impact, breaking up into small fragments. When meteors hit aluminum surfaces they vaporize on impact leaving a small dent.

"We test everything to failure," says Bigelow. "I am not a fan of computer modeling as being the dictate. We have done hypervelocity impact tests and the ballistic properties are fantastic, superior to the aluminum cans that are the alternative."

Bigelow's future space station complex showing large inflatables with connector nodes and docking ports. *(Images courtesy of Bigelow Aerospace Corporation)*

These inflatables actually have their roots in NASA's ISS development. They originated from NASA's Transhab module, designed by Dr. William Schneider in 1997 while he was working as Senior Engineer at NASA's Johnson Space Center. Transhab was and inflatable cabin that was originally intended as a manned habitat on Mars, but was later configured to house a large crew of astronauts aboard the ISS. A roomy lightweight module was to test the feasibility of fabrics in space instead of the heavy metal segments that are now standard for space colony designs. The Transhab was to house between 8-10 people to live and work on the ISS, and still protect them from the harsh gale of radiation, meteors, and extreme temperatures. The project was cancelled as part of the first wave of budget cuts in the early George W. Bush administration.

But in 2006, it became a startling example of the available spin-off technology that can be cleverly utilized by entrepreneurs. In the initial stages, however, Bigelow has to depend on foreign countries for launch services, and has run into government regulations prohibiting or restricting their use. The obstacle is the new export control regulation, International Traffic in Arms Regulations (ITAR).

Conceived to restrict the flow of technology to and from foreign powers, the regulation is slowing the process of space business internationally. Over the past few decades, there has been great concern over dual-use aerospace technology that terrorist organizations might acquire. Space entrepreneurs are particularly vulnerable to this kind of restriction, and should be given extra consideration, given their tenuous funding and limited domestic alternatives.

"ITAR is a very serious problem," says Bigelow. "I'm not saying it's all bad, but it needs to be trimmed back"

Free foreign trade in space hardware is essential for keeping the prices down. Without diversity in equipment procurement, the commercial space age would slow considerable. The main reason that rockets like Dnepr exist is because alternative U.S. vehicles are more than twice the cost. ITAR places restrictions on how easy it is to buy and sell equipment internationally. This would have to change (within reason) to accelerate space commercialism into the global marketplace.

The Genesis 1 also carries science missions. The payload includes Madagascar Hissing Cockroaches, to test their endurance in the tenuous air. Also aboard are Mexican jumping beans (caterpillars encased shrub seeds) to see if they germinate into butterflies. The pleasant, 79 F. degree temperature inside the spacecraft should provide a nice home for these critters until they inevitably expire. Also included are a variety of photos, business cards, and toys belonging to company employees, including Spongebob Squarepants and Superman dolls.

The most significant instrument aboard is an automated cell growth laboratory called Genebox. Developed by Ames Research Center for $1 million, this shoebox-sized lab is designed to conduct cutting edge genomic microgravity science and send data results back down to Earth.

Though only a calibration mission, Genebox is a direct result of NASA becoming a customer to private industry in lieu of microgravity science opportunities aboard the ISS. Operating on 2-4 watts of electricity, Genebox monitors a number of its internal parameters including system timers, control set points, status indicators, temperatures, payload pressure, currents, voltages, relative humidity, radiation, vibration, fluorescence, and optical density data, and reference readings from optical system components. Bigelow Aerospace arranged to have the lab flown free of charge and is interested in future arrangement with NASA's Ames Research Center.

The Genebox payload is an example of affordable telerobotic instruments aboard commercial space stations where the payload only sends back data and not the actual samples. If in the future, telerobotic analyzers and even pilot production facilities could do the same, the cost savings would be proportionate. The taxpayers will save money if NASA uses accessible commercial facilities on an as needed basis.

Bigelow's business model contains one big gamble: funding the launch costs. Despite the $9.5 million cost of the Dnepr, this is only a one-shot deployment flight. A manned flight would need to use a much heavier launch vehicle that can come back down safely. There

The Genebox, tested aboard Genesis 1, will analyze how microgravity affects genes in cells and other small life forms. *(Image courtesy of NASA)*

are a few options Bigelow may use, but none have a crew-rated launch system in service in 2007. In fact, Bigelow has no way to affordably servicing or supplying a space station either. He is literally waiting for a suitable spacecraft to come along soon.

To hedge this huge gamble, Robert Bigelow has tackled the problem on a few fronts. One of which is to invest in a new, affordable launch vehicle by volunteering as an early customer. Much like Rene Anselmo did with the new Ariane rocket, Bigelow could gamble on one of the maiden flights of an experimental vehicle. His choice is the Falcon 9, a simplified, reusable booster made by SpaceX Corporation. This company, based in El Segundo, California has won NASA support for re-supply the ISS, but is struggling to get off a successful launch. On its maiden flight, the rocket failed. The second flight was only partially successful.

Another possibility is Lockheed Martin's Altas V vehicle: a crew-rated version of the current satellite launcher, complete with emergency abort capability. Atlas could be outfitted for a complete low Earth orbit transportation system. It has the power, the support, and a nearly immaculate flight record to back it up. All that's needed is the funding. Though the Atlas Vs are perhaps the best rocket available, they cost approximately $80 million to launch. Though far too expensive for Bigelow Aerospace, a study is currently going on to find out if the Atlas can be made as practical as hoped.

In lieu of a breakthrough, Bigelow Aerospace is sponsoring a contest to spur the development of a new orbital transportation system. They are prepared to award $50 million for the first entrepreneur to build and fly a spacecraft to low earth orbit.

Called America's Space Prize, it was offered to foster new concepts in space vehicles that go way beyond current sub-orbital attempts like the SpaceShipOne. Instead of going a mere 62 miles, the winner of the America's Space Prize will have to reach 250 miles, travel two times around the Earth, and land back on the ground. It must also do it all over again within two months. The spacecraft must be 80% reusable, must not be funded by government money in any way, and finally, must carry no less than five persons aboard. These requirements must then equate with low operating costs; low enough for Bigelow to use.

Commercial space travel is not for the faint of heart, whether you are flying to the stars, or driving to the bank. The best market, according Robert Bigelow is in foreign governments. In his interview with The Space Show in the summer of 2006, he said that many foreign astronauts will pay for the opportunity to go into space. Their respective governments can afford the high price, and since the return is not commercial, there is no need for a hyper-lucrative payload to justify the mission.

If all goes well with the Dnepr rocket, the next mission, *Genesis II*, which will use more cameras to target specific ground sites. Still an unmanned test vehicle, *Genesis II* was successfully launched in late June of 2007. The pathfinder vehicle sports 22 cameras, a multi-tank inflation system, and a bingo game.

A biology experiment called Biobox is also flown on Genesis II. The box is divided into three chambers for ant farms, cockroaches, and scorpions. These experiments are to examine insects in microgravity and to test life support systems for future manned missions. These insects are very hardy and should do fine with minimal life support systems. If they don't survive, humans cannot be flown until the problems are solved.

There is also an extensive "Fly Your Stuff" menagerie of personal paraphernalia aboard the Genesis II, including advertising media.

Bigelow plans bigger and better stations in the near future. The first phase of deployments is already in the works. The next model is called the *Galaxy* (55 cubic feet), with enhanced, life support controls, communications, and muli-tank inflation system, rather than the single tank on the Genesis 1. It will also have additional hull layers for added protection. The launch date is planned for the summer of 2007.

Next in line is a larger, 587 cubic foot (180 cubic meter) inflatable called *Sundance*. It will be able to house a crew of three astronauts, provided they can get a round trip to Earth aboard an Atlas V, Falcon 9 or similarly large rocket. It will have a mass of 8,618 kilograms (18960 lbs.) and be equipped with life support, attitude control systems, three windows, on-orbit maneuverability, re-boost and de-orbit capability.

The final phase currently on the books is the *BA 330*, or Bigelow Aerospace 330 cubic meter (1080 cubic foot) volume station. Also dubbed, the *Nautilus*, it will be 29 times the size of the Genesis 1, and will service the tourists and business scientists who will try to make the microgravity spacecraft pay off.

The different classes of inflatable modules.
(Image Courtesy of Bigelow Aerospace)

Present and Future Bigelow Aerospace Modules

Module	Interior Volume	Hardware
Genesis I (unmanned)	11.5 cubic meters	13 cameras Single-tank inflation system Solar powered communications.
Genesis II (unmanned)	11.5 cubic meters	22 cameras. Multi-tank inflation system. Enhanced reaction-wheel system. Enhanced communication capabilities Additional hull layers
Galaxy (unmanned)	16.7 cubic meters	Advanced onboard avionics. Key elements for testing the environmental control life support system. Upgraded attitude determination and control system. Multi-tank inflation system. Increased communications bandwidth. An improved, more robust air-barrier. Solar and battery power improvements. Stiffer fabric structure.
Sundancer (manned)	180 cubic meters	Not yet available
BA 330 (manned)	330 cubic meters	Not yet available

APEX – Spacehab's Customized Retrievable Spacecraft

In great contrast to Bigelow's inflatables, Spacehab, a grizzled veteran in the commercial space wars, has a more traditional approach: a free flying laboratory that doubles as an ISS re-supply ship. It can be either manned or unmanned, and can be retrieved so scientists can get back products created in space. It can be requisitioned by the government, or by private industry. A pretty safe marketing bet, but still a formidable technical and financial challenge.

APEX. Spacehab's re-configurable space vehicle.
(Image courtesy of Spacehab, Inc)

While talking with Mike Bain, then the Chief Operating Officer of Spacehab, he made clear that APEX is only an architectural concept at this time, but its versatility demands the attention of those interested in sending up experiments and getting them back after a short period of time in space. The retrievable capability and short turn around time of the APEX fills the gap left by Space Island Group and Bigelow. The business angle of APEX is in its versatility. It can be designed to do anything you want, and many experiments can be flown on the same flight.

"The APEX is best described as a Mission Re-configurable Modular Spacecraft," says Mike Bain. "We think that with this vehicle

[we] can do many things at once, and by doing that, we can put all the customers on one mission and reduce the cost of the operations."

APEX is configurable, and can be made available in different sizes. There are one, three and four meter long versions and each can be compatible with different applications. The launch vehicles would be variable as well. An APEX could ride on top of either a large Atlas/Centaur booster, or a small Minotaur missile depending on the payload. After it's time in space, the APEX would splash down in water and be retrieved by boat.

Since 1984, Spacehab has provided NASA and commercial space companies with habitats, stowage containers, cargo carriers, and laboratory racks. The company has had a long relationship with NASA and the Johnson Space Flight Center in Houston where they create full-scale mockups and training equipment used in the Shuttle and International Space Station programs. Even today, Spacehab racks, or pallets that hold hardware, are used to strap down equipment for the ISS. The company offers start-to-finish space flight services in three categories:

1. Flight Services – vehicles, containers, cargo carriers.
2. Launch Processing – contamination free encapsulating, stabilizing payloads for launch to survive the vibration and acceleration of launch and the vacuum of space.
3. Human support in hardware design, fabrication, mock-up, safety, quality assurance, and data management or payload.

The APEX concept replaces the now shelved project known as Enterprise. In the early 2000s before NASA shifted its science goals to aid exploration, Enterprise was a candidate for a space module devoted to marketing space activities. It was supposed to house media projects for educational television stations, such as the Discovery and Science cable channels. Enterprise was to be a joint project between the Russian prime contractor RSC Energia and Spacehab, and was to be launched by the Space Shuttle and attached to the ISS. Its two decks were to house commercial payloads that would show the public the phenomena of microgravity.

The APEX series of re-configurable modules for broad market access.
Spacehab hopes to accommodate multiple payloads on each flight.
(Image courtesy of Spacehab, Inc.)

As the recession wore on, the project languished. Then a new market emerged for supply and retrieval of ISS payloads: COTS (Commercial Orbital Transfer Service). This was NASA's request for a commercial carrier to step forward and produce an orbital spacecraft. APEX entered the contest but was turned down for funding. The market for free floating space stations; however, is in the forefront of the minds of people who know what they can uniquely provide.

"We think there is still a market for APEX, says Bain. "We bring some capabilities to the table that they [the competition] don't think NASA really needs. We can also use it as a microgravity lab coming and going from the station."

Free-floating space platforms are not associated with any mother vessel. They sail through orbit without the vibrations and accelerations that large multi-purpose space stations produce. They can be manned or unmanned, and can have many different uses such as:

1.   Retrievable cargo carrier for ISS

2. Experimental platform for long duration tissue growth, crystal growth for jewelry, optical lenses, semiconductors, and electrical component plating.
3. Orbital platform for science missions such as telescopes, radio payloads.
4. Exposure facility for components that need to be tested before they fly in space for long duration.
5. Advertising, memorials, artifacts, branding platform.

With accessible space stations in full operation, solutions to some of today's most tenacious problems, like medical treatments and energy sources, can be duly addressed. To make these new habitats affordable, mass production must be available and a wide market must be tapped. Most importantly, the keys to the next conquest of space is identifying the best markets and developing affordable applications to exploit them. Only then can the common citizen truly benefit from the space frontier.

4

MICROGRAVITY'S RAINBOW:
DEFINING MARKETS IN LOW EARTH ORBIT

I N THE YEAR 1800, IF you knew that the railroads would crisscross the United States from Boston to Sacramento, you would invest in steel. In 1900, if you knew that automobiles would run on gasoline for the next 100 years, you would invest in oil. In 1950, if you knew roads would be built to expand America's byways from Miami to Anchorage, you would invest in commercial real estate. During the American westward expansion, farming and mineral riches glinted on the horizon. Today, markets in near Earth orbit are coming into view and may be the drivers of the next economy.

If the microgravity economy is ready to expand by 2010, we will need safe, affordable, reliable infrastructure, both in transportation and in habitation before much of the investment dollars can be expected. Understanding the major space markets is essential.

The most important fundamental basis for doing anything commercial in low Earth orbit is defining markets as carefully and as thoroughly as you can. Although they surely exist, the practical method of exploiting of them does not. It all boils down to one proportion that is obvious to some, but seems elusive to others. It is a fundamental constant in the pursuit of affordable access to space:

The profit generating capacity of the payload must exceed the cost of the launch, operation or re-supply.

Space Island Group's multi-decked microgravity laboratory. A turnkey space station outfitted to continue microgravity research when the ISS is decommissioned. *(Image Courtesy of Space Island Group)*

This constant is independent of the launch vehicles themselves because all is proportionate. If the payload is a mining facility to extract half a billion dollars out of a near Earth asteroid, a $200 million dollar launch budget is certainly acceptable. If the payload is a tiny satellite made for a university that depends on limited resources to deploy, even a launch cost of $10 million is prohibitive.

Many folks get preoccupied with vehicles that will provide affordable access to space. A scramble for low priced spacecraft is underway and should be, but large communication satellites have always produced revenues and profits even using the most expensive boosters available. Profits primarily depend on making the space markets as lucrative as they can be, *and on schedule*, with or without the means of reducing the cost of getting there. Expanding orbital markets is the key.

The space market is vertical, even with today's expensive rockets and high insurance costs. Today' killer applications in near Earth orbit are communications satellites, tomorrow's will probably be tourism and medical research. The things to watch for are new developments in space transportation systems that do not physically or legally restrict the flow of passengers or products to and from low Earth orbit. The cost/benefit ratio must be very high to exploit markets in space. Lowering costs for launches and insurance are always positive, but not essential.

Finally, I will not get grandiose here. Microgravity can improve products and processes that we can do on Earth, but because of the extra effort involved in getting into space, the improvement must be revolutionary. There really isn't any reason to go into space if the improvement is only slight or moderate. As far as saving the planet through solar energy -- cool it! Investors must safeguard themselves from bankruptcy before they can start cleaning up the atmosphere. The major space markets described here show clear potential for large revenues, though actual dollar amounts are still too vague to honestly quantify.

Safety is obviously paramount as well, both to personnel and equipment. Without safety at a fair price, there cannot be an inhabited space commercial world. NASA and Roscosmos have descent track records. America has lost seventeen astronauts and Russia, four. This is after 45 years of nearly constant manned space flight. Can the commercial world create the same safety record without self-destructing? It remains to be seen. One thing is certain, many will die and many assets will be lost, but this danger will never change people's minds about this new world any more that it did about Alaska during the Klondike Gold rush in 1897. If there is wealth to be attained, there will be those who venture into the forbidding frontier to find it.

5

SOLAR POWER SATELLITES:
TAPPING THE SUN'S LIMITLESS RESOURCES

S INCE THE FIRST CONCEPTS IN 1960s, solar power satellites have been touted as a pollution-free alternate to nuclear and coal burning power plants. The designs have been around for years and are basically simple: create a large expanse of panels at an altitude of 23,500 miles to collect the sunlight, and beam down the solar energy to the Earth in the form a microwaves. Its geosychronous orbit allows it gather sunlight almost constantly throughout the year, and the microwaves it sends to Earth will penetrate thick clouds. These facts make solar power satellites superior to ground based solar arrays.

On the ground, the receiver, or rectenna as it is called, is a large field of wires that conducts the beams of microwaves energy and puts the juice into the public electrical grid. Up to five billion watts per satellite could be generated without appreciable damage to the local environment. By comparison, the largest nuclear plant in the U.S., Palo Verde outside of Phoenix, Arizona has a 3.7 billion watts capacity. Grand Coolie Dam in Washington State, the most productive hydroelectric plant in the U.S., generates 7 billion watts, or seven thousand million watts! It is often believed solar power satellites will save our planet from global warming and be the prime candidate for clean energy sources in the near future.

These highly efficient solar satellites, each covering one square mile of space, could be deployed to help developing nations meet their high energy demands. *(Image Courtesy of Space Island Group)*

But in reality, this will probably never be the case. The largest solar satellite ever conceived: a "Manhattan Project" colossus, spanning three by six miles in size, could supply more power than the largest nuclear plant in the U.S., depending on how efficient the solar cells are. But even at 37.3 % efficiency using the latest in gallium indium arsenide solar cells, boosted by magnifying lenses and superconductors to increase their power, the world would require an armada of these to meet the world's energy requirements, which is about nine to twelve trillion watts. This is a daunting prospect to contemplate. What is much more feasible is to build smaller versions to make profits for localized energy companies.

The solar power satellite concept, first conceived by Dr. Peter Glaser in 1968, comes back into vogue every time there is a mideast conflict and an energy crisis. (Iraq is our 4th). As time goes by, the idea becomes more political and less technical. NASA and the Department of Energy in the late 1970s extensively studied a twenty-square-mile, five billion watt solar satellite. The conclusion was that the costs of launch and maintenance were prohibitive. The project was shelved because of the perceived level of difficulty of attaining financial resources, and building the satellites themselves. Launch costs were the biggest spoiler of all, since the Shuttle is $400 million a flight.

The reason why the idea will not go away is very clear. Unlike ground based solar, satellites can remain in sunlight 99% of the time allowing for constant power generation day and night. (There would be a blackout of 75 minutes per year at the beginning of spring and fall). The satellites would be in geostationary orbits that would always float above certain landmasses. In these high orbits, there are no known disruptive perturbations engineers can't handle. There is a thin rain of meteors, but solar panels can be made sturdy enough to withstand this. Radiation resistance can be built into the cells as well. Today's communications satellites, using these very same kinds of solar panels, can last 20 years in high Earth orbit without a serious malfunction.

No government in the world currently has plans to build such a system. However, with energy as the No. 1 priority in an ever growing, developing world, the solar power satellite solution cannot be ignored, particularly with India and China on such a fast track in their economic expansion efforts.

Solar power from space has the advantage of not needing any transportation fuel to operate. Coal and nuclear power plants require supplies to be shipped continuously to and from the plants. They are ongoing expenses for residents and a hazard to the environment when spent nuclear fuel or smoke is created. Solar power satellites are investments that will keep on giving for about 40 years and will emit no waste at all.

Microwaves can be beamed to certain areas of the world, providing power to the lucrative locations. By steering the transmitting antenna, solar satellites can supply the power grid to various locations where rectennas are built. Instead of operating coal burning plants or hydroelectric dams everywhere, the environmentally safe solar power beaming system could provide energy to rapidly developing areas of the world that are hardest hit by fuel costs and pollution, like China or India.

In the War on Terror, solar power satellites are hard to attack. Unlike nuclear plants that are clearly vulnerable, solar spacecraft will be 23,500 miles high and almost out of reach of weapons terrorist usually get their hands on. Being almost one-tenth of the way to the Moon, the satellites would require terrorists to use lasers or modified ballistic missiles to punch a hole in the array. Though all these points have been made for many years, world governments, including the U.S. are reluctant to pay for even a small test of a solar power satellite. There is little doubt that the U.S. has to participate in some way to make space solar power a reality, but the perceived level of difficulty has never been surmounted, no matter who testifies before Congress in favor of it.

In October 1997, John Mankins, then NASA's Manager of Advanced Concepts Studies, researching new versions of solar power satellites, outlined before the House Subcommittee on Space and Aeronautics, a lengthy explanation called "Solar Power: A Fresh Look". In it he told of an updated system that used superconductors to enhance the power, thereby allowing satellite to be smaller and easier to launch. Low, medium and high altitude satellites were discussed.

Launch cost was the spoiler again. It was concluded that high launch costs were considered the barriers to any further development of the system. Vehicles needed to launch material at $180 per pound to make solar power satellites affordable. Current unmanned vehicles are approximately $3000 dollars per pound. Unless adequate launch vehicles were developed, the project would be too expensive. Though obviously necessary for U.S. to pursue, the "Fresh Look" idea was shelved by the U.S Congress.

Ralph Nansen was engineer from Boeing who worked on the Apollo and Shuttle programs, and led the late-1970s NASA-DOE solar power study. On September 7, 2000, he testified before the House Science and Technology Committee in favor of the concept.

His case was not to impress the Congress on new technical details, but to champion the common sense nature of the idea. He recommended a ground test facility be built costing $90 million to test the rectenna (receiver) technology. In his 1995 book, *Sun Power*, Nansen expressed concern that the nuclear industry did not appreciate the competition for the electricity market, and had more influence over Congress than the solar power satellites pundits. The lack of serious action from Congress on space solar power has become deafening.

In 1995, it was clear the Shuttle was too expensive for the job, so Nansen expressed the need for cheap, reusable spacecraft that could send up small, unmanned telerobots to put the satellites together. Nansen, like most other solar power satellites advocates over the years, hinged their proposals on low-cost-high-return launch vehicles. The problem is that no such vehicles are available from NASA or the commercial sector. Waiting around for this low cost vehicle has cost us 10 years of development time.

Even if we had low cost launchers, they might not be the answer. Automated spacecraft would have to go up thousands of times into low Earth orbit to build the array. Once finished, the satellites would have to be lifted into high Earth orbit by electric "ion" engines. No vehicles built today can carry out such tasks. They would have to be derived from existing space hardware or built from scratch. But if this hurdle could be surmounted, the business climate might warm to this pollution free, politically popular approach to energy generation. Private companies could justify developing dedicated vehicles to build the satellites.

The high-cost-high-return model works well for solar power satellites. It would require the use of expensive heavy lift vehicles taking up to 100,000 lbs or more into space at one time. If space stations were being built alongside the solar power stations, the cost could be defrayed. Owners and operators of the space stations would make their money off research, manufacturing and entertainment services. Energy companies would lease the space stations to build the satellites. Launches could be used for both projects. By helping each other in the process, space station investors and solar satellite investors would save money by combining their efforts.

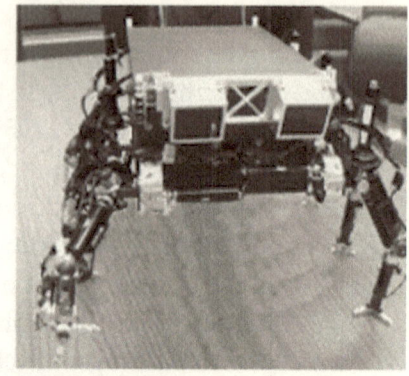

NASA telerobots may someday assemble solar power satellites, allowing astronauts to supervise operations. *(Photo courtesy of NASA).*

Gene Meyers, or Space Island Group, has championed this idea for a potential market so big, it dwarfs all costs of commercializing space. His dual launch vehicle (See Chapter 3) could support a serious project in solar power. He is offering fixed price contracts to existing aerospace companies who can build the satellites. Space Island would require them to design and mass-produce this hardware by modifying existing, space-proven components. Meyers expects $3.5 billion in revenue after six years of complimentary use.

"The 800 pound financing gorilla [for the space stations] is solar power satellites, says Meyers. "The big challenge is that 95% of the cost of building the solar satellites are the launch costs."

Using a smaller, more conservative model for solar power satellites, Space Island Group believes that a very high capacity, $1 billion factory could mass-produce a solar satellite one square mile in size. The conservative estimate for building the satellites would be $1,000 per kilowatt. Once deployed, the satellite could sell the electricity for about $0.10 per kilowatt-hour. The customers would be those who need electrical power in distinct areas of the world where rapid expansion is taking place.

India would be of the best target market. This country needs instant electrical infrastructure capable of leapfrogging the usual construction costs and liabilities inherent operating fossil fuel or nuclear electrical plants. India is the perfect frontier for solar energy. According to Space Island Group, 45% of all households in India do not have direct access to national or state grids. India's department

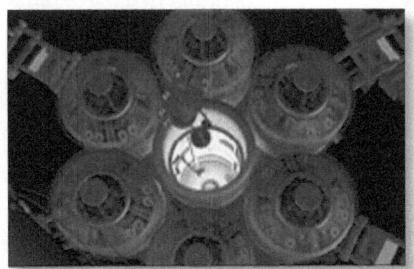

Six microgravity space stations joined in a cluster. *(Image courtesy of Space Island Group)*

Beneath the cluster are the robotic assemblers that build the solar power satellites. *(Image courtesy of Space Island Group)*

of energy plans to spend $300 billion in power generation by 2020. Infrastructure to distribute the power would cost another $250 billion. If the infrastructure is build, a 250% increase in electrical demand will occur in India alone by 2030.

Gene Meyers says the financing process would work like this: first a Power Purchase Agreement would be made with an energy company, which means a certain amount of electricity would be promised to a certain energy company over a long period of time, perhaps 20 years. A financial institution could guarantee loans for the electric company to build the solar power satellites. Meyers believes that this seed money could be used to build space stations and his launch vehicles. All launch costs could be recovered if the space stations pay for themselves in five years. Energy companies would only pay for the satellites and not for the cost of building them.

- Space Station Leases: $25/cubic foot per day
- Space Solar Power: 10 cents per kWh
- By 2020: 33 million cubic feet of station lease space and 18,000 MW of SSP capacity

Since every low-cost-high-revenue attempt has failed to convince investors, this fresh financial idea seems interesting. After a 40-year lifespan, the solar power satellites would generate trillions of dollars in revenue. There are no supply chains, hazardous storage areas, or government inspectors making the solar power satellites expensive

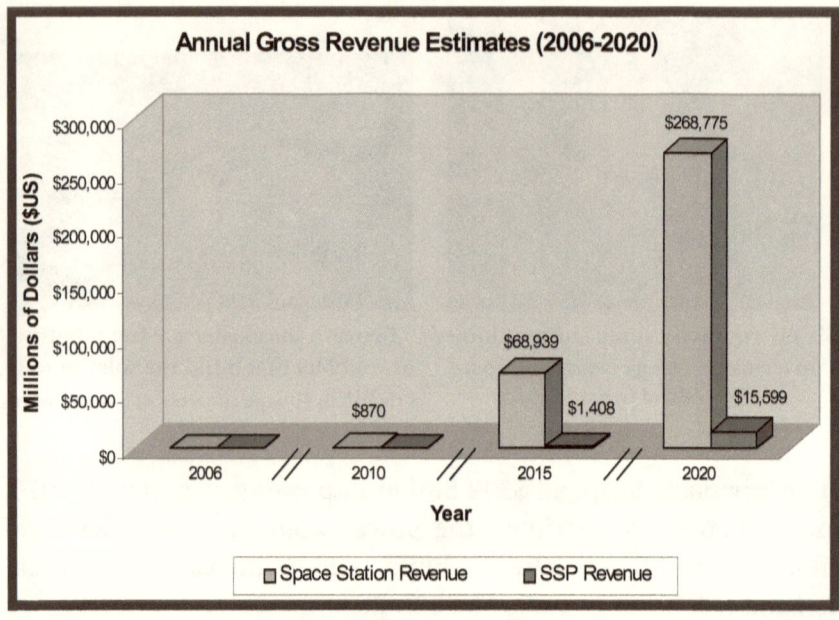

to own. If fact, the whole operation could be sold many times over before it needs to be decommissioned.

Here the high return eclipses the expense of reaching space. Low cost launch vehicles are not needed at all. In the meantime, as the winds of war chafe at our budgets, and global temperatures rise ever higher, the solar power satellites idea is still perceived as uncompetitive and costly. But for investors, the market base is clear: people need as much green energy as they can produce and are ready to pay if the rates are competitive.

6

PROTEIN CRYSTAL GROWTH:
DECIPHERING THE NATURE OF DISEASE

OUR BODIES ARE MADE OF proteins. There are in the cells of our skin, organs and hair. When you look into the mirror, protein looks back at you. There are over 300,000 kinds of proteins in the human body. They are indeed the "stuff" of life. Diseases attack proteins, affecting with way cells behave. Viruses themselves are DNA molecules wrapped in proteins. By understanding how proteins work, is to understand how to kill viruses and save human tissue. In effect, treatments and vaccines for diseases, from cystic fibrosis to AIDS can be developed by first studying the proteins of either the tissues or viruses themselves. Drugs are made kill or protect specific proteins.

Proteins, in the form of crystals are grown in liquids. Under the microscope, they look like bundles of rock salt floating in water. Biologists can grow them, molecule by molecule, in just a drop of solution. Once you grow them large enough, they can be imaged in fine detail by using of x-rays. If a large, well-structured crystal is imaged with an x-ray crystallography device, the crystal can be mapped down to its very atomic structure. Biologist can then find out what will protect the protein from disease, or the case of a virus, destroy or inhibit the functioning of the protein. Today's AIDS treatment drugs began with ground based protein crystal research.

Protein crystals do not grow very well on Earth. The problem is that the solutions crystals grow in are constantly swirling around. This is because of convection, or the tendency for liquids to move up and down in heat currents. It's like trying to build sand castles to close to the shore. Different ingredients in the solution have different weights, and they create tiny eddies in the protein crystal solution. The crystals end up being small and misshapen in most cases. The result is a 6-10% success rate in Earth based crystal growth experiments. In space, however, things are quite different.

The structure of purine nucleoside phosphorylase (PNP) for cancer drug research. *(Image Courtesy of NASA)*

This first experiment in protein crystal growth was conducted the Space Shuttle's United State Microgravity Lab One mission in 1984. Crystal growth experiments have been routinely performed in space ever since. Because there is no gravity in orbit, there are no convection currents, so the crystals will form without sheer forces upsetting the slow accumulation of molecules. The results are much larger and well-structured crystals. In space, 50%-100% of the crystals will form better, and some of them will be much larger and more articulated than Earth made crystals.

Crystals of recombinant human insulin grown in microgravity (left) are visibly larger than those grown at a ground-based laboratory (right, shown at the same magnification).

The space-grown crystals also proved to be better structured and optically clearer, making them more desirable for research. *(Images courtesy of NASA)*

Vapor diffusion is the popular method of crystal growth in space. Basically, it means that a biologist places a solution containing proteins in a device that evaporates the water solution away leaving crystals of protein to condense. Though this can be done in a short period of time, it is best to have a stable, 1-2 month's residence time in microgravity. Jitters in the spacecraft have to be minimized as well so as not to upset the formation process.

Protein crystal growth is the killer biological application. It is also the most popular experiment ever flown in space. Scientists from Russia, Japan, Europe and Canada on MIR or the ISS have performed protein growth for years now, and have produce high quality structures. There have been scores of protein crystal experiments on the Space Shuttle missions. The ISS runs protein crystal growth experiments today. The European Space Agency conducts experiments in their Protein Crystal Growth Monitoring by Digital Holographic Microscope (PromISS). With this device, growth and analysis can be conducted in the same operation. The combining of the newest information technology with long duration space flight can continue research at an optimal level.

Among the many proteins grown over the years are: Iysozyme, a protein from hen egg white that can be used to compare the quality of ground- and space-grown crystals; bacterial purine nucleoside phosphorylase (PNP), used for synthesis of anticancer drugs; human C-reactive protein (CRP), a major component of the human immune system; human serum albumin, and canavalin and concanavalin B, to improve the nutritional value of food sources and HIV Reverse Transcriptase Complex, an important protein used in the design of drugs used to fight AIDS.

One of the gurus of space based protein crystal growth is Dr. Lawrence DeLucus, the Director of the Center for Biophysical Sciences and Engineering at the University of Alabama at Birmingham. He flew in the Space Shuttle in 1992 where he performed 31 crystal growth experiments. He was so impressed by the results, he has been lobbying to Congress and the public alike to fund more growth experiments. In an interview on The Space Show in August 19 of 2006, DeLucus said that on Earth, nine out of ten crystals fail in the growth process. But in space the results are much better.

"If you fly it five times, the success rate of producing crystals, better than the best rate ever produced on Earth, is upward of 50% percent or higher."

A commercial long duration space station, manned or unmanned is needed to continue the research. The shuttles could only support the experiments for only 14 days, so there is not a lot of opportunity for pharmaceutical research. Because of the short duration of Shuttle flights, and the technical problems on the MIR missions, the process is could be considered as a competitive analytical technique compared to ground based versions.

"Protein crystals on Earth generally need a week to two weeks to grow to their full size," says DeLucus. "In space, [the] length of time that they need is actually much longer."

The funding for ISS biological science missions is now greatly reduced for the U.S., which makes a commercial space facility a necessity for protein crystal growth and other biological experiments. University, government, and commercial research money is waiting if a stable platform with enough power and reciprocity capability that can be deployed simply and profitably.

"A commercial enterprise that can guarantee a flight every month would be very valuable," says DeLucus.

The good thing about the crystal growth process is that it does not require astronauts to attend the experiments. It can be automated on a retrievable unmanned space station like the Long Duration Exposure Facility sent up in 1984. Dr. DeLucus and his research team have developed a way to perform crystal growth experiments in space using an automated device. It builds the protein crystal, and performs the x-ray imaging in orbit. DeLucus explained that all that is needed is to grow the proteins, take an x-ray image of each, find the ones that are best, and freeze them in liquid nitrogen.

Once in a protected solid form, the proteins can be sent down to Earth in a capsule for further analysis. There is even a partially competed apparatus, the size of a refrigerator, he and his team developed to automate most of the functions, but the money for space research ran out in 1999 so its development was shelved. The University of Alabama has on-going projects involved in space protein crystal growth, but the funding dried up for progress to advance further in space. An array of devices is warehoused waiting for implementation.

For a space protein crystal grown facility to be economically viable, frequent and reliable flights are necessary. Companies would need quick turnaround times.

"There were twelve proteins flown in space more than ten times," says DeLucus. "The success rate for those proteins were 100%. Probably, the ideal [residence] time in space is six weeks."

The next logical step is to build a facility onboard a commercial space station. The X-Ray Crystallography Facility, or "beam line" as it is called, would have to be pressurized and tended by a crew. Piloted flights to test the process would have to be made ahead of time. Onboard analysis from top to bottom is one of the benefits of space stations. This too would be a boon to research in many different biological areas.

NASA had proposed an X-Ray Crystallography Facility (XCF) for the International Space Station to support the preparation of crystals for visual evaluation and mounting, sample freezing, and collection of X-ray diffraction data. Without any crew intervention, the proposed system robotically downlinks video data to the awaiting scientist. The researcher then determines which crystals are best for

X-ray diffraction analysis and freezing. The downlinking allows the scientist to review the results and uplink any modifications to the data collection process. The current disposition of this apparatus seems up in the air.

The commercial protein crystal growth facility could be one of the most important research tools ever placed in microgravity. It can literally save lives and make millions of dollars for the first companies that can find treatments and vaccines for diseases like SARS or the Avian Flu. It can hasten the development of drugs used for the diseases that are plaguing society right now; such as, diabetes and influenza, and can be the antidote for every question on whether or not space science is worth utilizing. If there is a reason to go into space, protein crystal growth is it, whether you are a scientist, investor, or biotech firm board member.

A turnkey crystal growth process seems tailor-made for the first commercial space station. This refrigerator-sized device can fit easily into payload bays of today's rocket vehicles. Unmanned capsules, much like those the Russians and Chinese have sent up over the years, have at least the dimensions and the electrical power to run an automated protein crystal growth device on-orbit and bring it right back down again. The problem, as always, is in the cost benefit analysis per flight. The spacecraft supporting the protein crystal growth apparatus requires quick return of both products and investment dollars.

Should the American government help a commercial protein crystal growth facility on-orbit? The steady stream of diseases coming our way, and their impact on coming health care costs makes it hard for it to afford not to. With wealthy Baby Boomers entering the elderly periods of their lives, the investment possibilities become clear.

7

CELL GROWTH FOR ANTIBIOTICS AND TISSUE GENERATION: NEW TREATMENTS FOR ANCIENT SCOURGES

THE SUCCESSFUL DEVELOPMENT OF BIOPROCESSING in space is one of the seminal events in space science history. This is because microorganisms, foundations of biotech research, grow abundantly the peaceful environment of microgravity. After their discovery aboard the Biosat II research satellite experiments in 1967, accelerated growth of molds and bacteria in spacecraft have been both alarming and compelling to researchers.

One person's burden is another person's asset. Molds can be a danger to astronauts who might get sick from them, or it can be a blessing for the drug manufacturers who produce antibiotics form molds. Over the subsequent years of manned and unmanned space experiments, there has been a keen interest in utilizing space grown bacteria for making unique pharmaceuticals. Recent results on the ISS have included a 75%-200% improvement in antibiotic production over Earth based processing.

Most antibiotics come from animals, plants, molds and bacteria. Moldy bread creates a bi-product that kills bacterial infections in the human body. It is called penicillin. Eating moldy bread won't cure your sickness, but isolating the metabolite and creating an antibiotic from it will. Many antibiotics are man-made, but most are made from natural fermentation of bacteria and plants that will create "children" chemicals (metabolites). Sometimes second or third generation metabolites create the antibiotics.

Though still under scientific debate, most evidence indicates that microgravity provides an ideal environment for growing bacteria. The main reason for this is the low sheer environment of the water that the bacteria are grown in. While the solution is suspended in space, there are no eddies or currents in the water due to buoyancy, sedimentation or convection. This allows the cell walls of the bacteria to absorb and excrete nutrients without interference. It also allows cells to adapt very quickly to their new environment, helping them get started with their reproduction. Space is indeed the natural environment for bacterial growth. Earth's gravity actually inhibits it somewhat.

The benefits of microgravity culturing of bacteria are compelling. If the yields of cell growth can be increased, the production of new and existing antibiotics can be increased as well. The antibiotics industry is enormous. The current market is about $10 billion per year.

The financial frontier for antibiotics is unending. Infectious diseases, like all living things, will evolve and mutate, so that today's antibiotics will not always be able to cure them. New tools, like microgravity labs, need to be utilized in the continuous fight against existing diseases that evolve into new strains. Scourge diseases like Tuberculosis, Cholera, Typhus, Syphilis, Scarlet Fever, Plague, etc., will not stay stable in the natural environment, but will evolve resistance to antibiotics. The most efficient tools available are needed to stop these diseases, and stop new scourges that we do not yet know about.

This has prompted NASA, Roscosmos, ESA and JAXA and CNSA to test biomedical equipment in the Space Shuttle, MIR and the ISS research environments. The most prevalent is the work done on equipment made by BioServe Space Technologies, now a non-profit, NASA-funded provider of biotech equipment for the ISS. They produce a line of multipurpose bioreactors, freezers, and incubators used for the self-contained culturing or antibiotics in microgravity.

Stephanie Countryman, the business contact at Bioserve, says the equipment is available for use on commercial stations.

"All of our hardware is still available for commercial outlets to utilize if they so desire," says Countryman, "We currently have two of our CGBA units on board the space station. Currently, it appears these units will remain operable on the station for quite some time. These are available for commercial customers as well as the science community."

Over the years, the Space Shuttle and the ISS have produced a long list of bacterial cultures, too many to mention here. The feasibility to produce antibiotics was been proven, but the time spent on development was woefully inadequate to the need. Some of the most promising studies were sponsored by Bristol Myers-Squibb Pharmaceutical Research Institute in the mid 1990s. On Space Shuttle STS-77 mission in 1996, gas permeable bags produced more antibiotic than did microbes on the ground. In one case, the improvement was 200%. On flight STS-95 in 1998, production of the chemotherapy drug Actinomycin D was 75% greater that an Earth sample.

Now the ISS has a bioreactor onboard that does not have the funding or the astronauts to operate it. The Bioserve's Commercial Generic Bioprocessing Apparatus, or CGBA, would have continued commercial biotech research had the focus not shifted to exploration. The current three-person crew barely has the time to keep the ISS operating, let alone do many bioprocessing experiments. Though some insect and plant studies still proceed on the ISS, the potential for bioprocessing cannot be approached with the current funding allowances.

Tissue generation, the building of skin, bones, and cartilage for repairing damaged body parts, can also be done more efficiently in microgravity. Like bacteria, skin and organs, are developed from cell growth. To make tissues, like new skin cells, bone cells, and cartilage require some kind of foundation over which to form properly. These foundations, or "scaffoldings" as they are called, are made from polymers that are compatible with the human body. Like vines growing on tree branches, the scaffoldings are used to grow the cells into their proper orientations during the course of incubation. In the three dimensional, quiescent environment of microgravity, organs may be built around scaffolding more efficiently.

This was first attempted in the Shuttle/MIR program in the late 1990s. Between September 1996 and January 1997, Freed and Vunjak-Novakovic with NASA colleagues grew cartilage aboard the Space Station Mir in the first tissue-engineering experiment in space. They published their results in the December 1997 issue of the *Proceedings of the National Academy of Sciences*. Though only a feasibility study, microgravity provided the 3-dimensional environment that is more like the inside of the human body in which the tissues actually grows.

This experiment has since enhanced Earth-based research. A new type of bioreactor was devised that simulates the microgravity setting. By slowly rotating the vessel filled with the tissue generating solution, a continuous state of free-fall can be recreated on Earth. Led by Dr. Lisa Freed, a principal research scientist in the Harvard-MIT Division of Health Sciences and Technology, working with Dr. Gordana Vunjak-Novakovic and other colleagues at MIT, Harvard Medical School, Boston University, and Brigham and Women's Hospital, heart tissues were successfully produced. Conceivably, an entire human heart could be created this way.

In Earth laboratories, much of the time and expense is spent to create conditions like those you already have in space. An orbiting space station, dedicated to biotech research for antibiotic and tissue generation, could help address the challenges of disease control, both scientifically and financially. No one can predict how important these new tools will be in the daunting century ahead, but we will surely need to put forth all the efforts we have available.

8

TOURISM:
EXPANDING THE ENVELOPES OF EXPERIENCE

THE SpaceShipOne, a suborbital spacecraft built by Scaled Composites Corporation, specifically designed for tourism, entered space for the first time on June 21st 2004. It was the beginning of space tourism as a completely civilian enterprise. Billions of dollars in revenue per year are possible in this growth industry. The market can only expand. Tourism recently has become has become a mature space enterprise, following in the footsteps of satellites and launch systems. As with most space innovations, its origin is in Russia. Its near future will be in the purview of the rich who have the time and fortunes for sub-orbital flights and visits to the International Space Station.

This is still *adventure* tourism, however. Deep sea diving and mountain climbing is more comparable to space tourism than beach combing or skiing. But what is being done to make it a middle class market that is affordable to many?

Space tourism broke onto the scene with Dennis Tito's historic flight in 2001. His space odyssey carried him to the International Space Station after the Mircorp Company could not sustain its lease on MIR. Space Adventures, Ltd. who dominates orbital tourism to this day, provided all the arrangements for the space cruise and training in Russia.

His weeklong stay was a shock to most people who expected NASA to keep him grounded. But the Russians, who were cash-strapped, insisted he go despite NASA's safety concerns. (It was never in the NASA charter to accommodate tourists in space flight and never will be). Since then there have been a succession of tourists:

Anousheh Ansari's spartan accommodations on her $20 million, week-long stay on the International Space Station in 2006. *(Image courtesy of NASA TV)*

encryption guru, Mark Shuttleworth, U.S. scientist and entrepreneur Gregory Olsen, Anousheh Ansari, founder of two telecom companies, and software developer, Charles Simonyi, all enjoyed a week on the ISS.

Though most of us civilians dream about traveling in space, it has become a reality for very few. These were very lucky, very wealthy folks who could spend six months away from their businesses to train in Russia. They were motivated and in good health. They could handle space adaptation syndrome and the frightening events that are inherent in the launch process. They learned how to use spacesuits and trained for unthinkable emergencies. Theirs was a perilous adventure, like deep sea diving along jagged reefs, or boating through Amazon swamps.

Future tourism in space has to be affordable. As discussed earlier, it must be comfortable as well.

Space Island Group has considered orbital tourism as essential to its revenue stream. Gene Meyers, the President and founder has always believed that bigger is better when it comes to space stations, but comfort should not be at minimum. This means that artificial gravity may be necessary for tourism to thrive beyond its beginning stages. Why? Most people, even some astronauts, have a hard time adjusting to weightlessness. Motion sickness is notoriously tough on tourists.

From the days of Yuri Gagarin, motion sickness in microgravity occurred in 60%-80% of astronauts. It is the first and most dramatic effect you feel in the first few days in orbit. Much like seasickness, it is caused by the vestibular system's disorientation when there is no point of reference for the balancing sensors in the inner ear. The body does not know where it is or how to balance itself. Most space travelers get over this is a few days; others never really recover during their visits. They need medicines and head braces and do whatever they can to not to upset their balance.

Contrary to the usually downplaying of the subject, there have been examples of extreme space sickness with civilian astronauts. The first civilian cosmonaut on MIR, Japanese journalist, Toyohiro Akiyama who flew back in 1990, was sick most of the seven days he spent in there. Jake Garn, a republican senator from Utah, a civilian astronaut in the Space Shuttle, became so sick, NASA created the scale in his name to measure motion sickness levels. On a scale of one to ten, Garn was a thirteen.

Space sickness is a malady even to the highly trained. Skylab astronauts, who had no indication that they would be sick on Earth, became so stricken that it directly impacted their busy work schedule. Pilot William R. Pogue, known as "Iron Belly", for his ability to adjust to low gravity forces, became stricken right after launch. Even though other pilots described him as having "cement in his inner ear", he took about a week to adjust to the Skylab environment. Because of the delay in getting their "space legs", the schedule was interrupted to the point where the astronauts could not catch up.

Motion sickness drugs help. They disconnect the inner ear sensors from the stomach. The remedy for Skylab astronauts was a readily available drug known as Scoplomine. The drug is also considered a hallucinogen and "truth serum" and is widely abused. Promethazine is another widely available drug is now being tested by NASA as a new motion sickness treatment. The side effect is drowsiness, so caffeine or some other stimulant needs to be taken with it. These drugs can only be used for a short period of time until the passengers get their "space legs". But what if they keep vomiting? What can you do about it in space?

The fact is that most of the astronauts flying in space are from the Navy or the Air Force. People who are just flying for fun may get sick more frequently. With the discomforts of the body fluids rising to the head, the stuffiness, dehydration, and sweating, there has to be better way to enjoy a trip to space, especially for children. The only solution is to counter microgravity with a place for people to normalize themselves in case they cannot handle the weightlessness.

Artist depiction of empty rocket boosters connected into ring shapes. Advanced space stations will be rotated to provide artificial gravity on the inside of the rim. *(Image courtesy of Space Island Group)*

Space Island Group envisions a ring of large empty fuel tanks that can be fitted with carpets, tables, view screens and dining rooms along the edges of a rotating wheel. Towards the center of the ring is the microgravity recreational area. As utopian as these sound, medical needs for some of the passengers may demand it. Customers need airy hotel rooms, at least 1/3 Earth's gravity, and toilets and bathtubs that work just like they do on Earth. Pioneer living is fun for a few days, but after that, tourists may just want to go home if they cannot have creature comforts.

Liquids float precariously in space. From body secretions to bath water, they are very difficult to control and clean up. In one story, a commander on a MIR flight was hurling through the space station only to come in contact head on with a basketball- sized blob of antifreeze that had leaked out of the cooling system. His hair skin and eyes were saturated with it. This was an exclamation point to any otherwise harrowing experience on the aging, decrepit MIR. In another floating fluid disaster, a backed up toilet left Space Shuttle astronauts with the daunting prospect of spending their "voyage of lifetime" dodging pieces of fecal matter.

Without artificial gravity, liquids do not stay put and can clog up ventilation filters. All sorts of hygienic debris from lotions, hair gels, shaving cream, toothpaste (and a lot of stuff I should not mention) will be floating around uncontrollably. There is a difference between trained pilots and tourists when it comes to tidiness. If commercialism wants to get past its "heroic" period, accommodation will be necessary.

What would be the dimensions of a space station that uses rotation to simulate gravity? Would there be discomfort from the rotation?

A large space station (1500 feet in diameter) provides artificial gravity by spinning rather briskly, about 2 times per minute for one gravity. A smaller station with a diameter of 500 feet would have to rotate once every three minutes. Gas impulse jets mounted on the outside of the station would provide the rotation by all firing in the same direction. The resulting artificial gravity would allow people to sip champagne, shower, recover slowly from their rigorous flights, and stay health and fit. The crewmembers will benefit greatly since they will have to stay for months at a time on-orbit.

One argument against rotating space stations is the Coriolis Effect, a state of relative motion described by Gaspard-Gustave de Coriolis in 1835. Pretend your standing on the inside of the rim of a rotating space station that is spinning fast enough for one "G" of force. Your head is pointing towards the hub. Your feet are squarely on the deck, pointing in the direction of space. Things may seem normal. Now start walking in the opposite direction of the spin. You feel something pulling you back. Now try walking sideways, perpendicular to the spin. You feel the side pull. You might feel like you are being thrown, not only to the floor, but also in the direction of motion. These are the two frames of reference your vestibular system

Luxury accommodations including flat screens will adorn the inside of a rotating space station. Stationary cameras will take live images of the Earth outside. *(Image courtesy of Space Island Group)*

gives you. Rotating the space station may cause as many problems as it solves. Only tests will tell for sure.

The fact is, nobody has yet tried to rotate space stations, even though it might be the key to our long duration existence in space. Developing artificial gravity is certainly easier than trying to stop bone loss or muscle atrophy. For today's astronauts and tourists who are well trained, microgravity is not a problem. For business people conducting industrial chemistry, weightlessness is even *desirable*, but to wealthy families on vacations, microgravity may seem like only an unpleasant part of space going experience. Maybe an aspect they would find interesting for a time, but too uncomfortable for a long stay, especially at those prices!

The well-to-do passenger is where the bulk of the tourist money will come from, at least in any sustainable way. Unless spaceships want to follow in the footsteps of the Concord supersonic jet, they must not only become affordable to the middle class, but sensitive to their expectations of comfort and convenience.

The tourist-friendly interior of a rotating space station.
(Image courtesy of Space Island Group)

And the cost? Space Island's model for high cost/ high return is the most feasible way to afford a vacation in space. If a family of three rents a cabin 25 x 25 x 10 foot cabin for one week, they would pay $1,093,750. It only gets higher with Bigelow's 330 or Spacehab's APEX. Apex will be $8,250,000 per person (plus supplies) for the large (26 cubic foot) module. This is still better than over $20,000,000 million for an ISS stay. If accommodations can be made luxurious and comfortable, more of the wealthy clientele will consider an adventure cruise in space.

Space tourism is adventure tourism for the wealthy, and the profit potential is coming clearer. In a survey performed by the Zogby International research firm, 450 wealthy people who made over $250,000 or more per year were asked if they would pay to be tourists in space. Twenty percent polled would pay $100,000 for a suborbital flight, but only seven percent would pay $20 million for a weeklong stay on the ISS. Somewhere in between is the tourist market for commercial space stations.

Tourism is the surest bet for the space entrepreneur, because as many have proclaimed over the years, there is no experience on Earth that can compare with it.

9

SATELLITE SALVAGE AND MAINTENANCE:
ON-ORBIT RECOVERY AND RECLAMATION

A $300 MILLION DOLLAR LOW EARTH orbiting constellation has suddenly gone silent. Engineers scramble to find problems with the software, orbits, transponders etc. while they hemorrhage revenues. Outraged customers, who have taken years to cultivate, are about to switch other providers. It turns out that a few of satellites have started to tumble, and that is causing the others to miss their relay links. The software to correct the problem has been corrupted. A scandal is brewing and a forfeit of business will result if the system cannot be bought under control.

Fortunately, there is help. Onboard a nearby commercial station, are automated telerobots that can retrieve the wayward satellite. Once aboard the space station, they can be examined by human attendants and adjusted. The telerobots can then place them back into their correct slots. Subscribers and insurance carriers are delighted.

The truth is, there are not on-orbit services for satellite maintenance. Operators cannot repair a satellite if it is malfunctioning or drifting from its orbit slot. All they can do is to adjust the software and hope for the best. Today there are about 300 viable satellites that are stranded or drifting. Most are just aging; some are catastrophically broken, losing their owners millions in uninsured revenues and lost science. These wayward satellites are opportunities for on-orbit salvage team to:

1. Re-adjust and repair satellites
2. Salvage or recycle satellite parts
3. Build new satellites and launch them from orbit.
4. Bring old satellites down for repair and future redeployment.
5. Perform post mortems on satellites to provide reasons for the failures.
6. Lower insurance cost for satellite companies.

The Space Shuttle conducted a few satellite rescue missions in the 1980s and 1990s. The most publicized was the Hubble mission that saved taxpayers billions, and gave fabulous images from space for the next decade and a half. Others were Intel's Westar-VI, and the Indonesian satellite Palapa-B2, both were launched by a previous Space Shuttle mission but malfunctioned soon thereafter. These had to be re-launched form Earth at later dates. (The Sattel Technologies Company did the historic refurbishments.) In another Shuttle mission, a stranded geosynchronous satellite Syncom IV-F3 was captured in low orbit after a kick motor failure. It was launched from the Shuttle while still in orbit and reinserted into a high geosychronous slot.

This was all part of the commercial applications program that gave way when Challenger exploded in 1987. (In 1992, another lost bird, Intelsat VI was sent on its correct path by Shuttle crews on-orbit, under a new Presidential space policy.) A huge gap in the market remains. Knowing the revenue potential of satellite rescue, Orbital Recovery Corporation, a London based firm, intends to build spacecraft to service satellites in stranded or degraded orbits.

Communication Satellites Repaired/ Retrieved by Space Shuttle Crews

Flight	Date	Satellite	Mission
STS-41C	11/84	PALAPA-B2	Returned to Earth and resold after not getting into proper orbit after STS-41B failed to deploy them correctly
STS-51A	11/84	Westar-VI	Returned to Earth and resold after not getting into proper orbit after STS-41B failed to deploy them correctly
STS-51I	9/85	Leasat/Syncom IV-F3	Correctly placed back into geosychronous orbit.
STS-49	5/92	INTELSAT-VI 603	Rescued and sent into proper orbit.

The rescue of Intelsat 6 by Shuttle astronauts aboard the Endeavor in 1992. *(Image courtesy of NASA)*

Three principle reasons exist for satellites to become disorientated.

1. Failure to deploy. The rocket engine malfunctions so that the satellite is in the wrong orbit from the beginning.
2. Depleted station keeping gas. The satellite runs through its attitude control gas during its expected lifetime and drifts into a degraded (inclined) orbit instead of a fixed geostationary orbit. This requires expensive antenna time to track satellites unnecessarily.
3. On-orbit mechanical failure for known or unknown reasons.

The engineers at Orbital Recovery plan to do this by deploying a spacecraft that attaches itself to a wayward satellite and sets it into to the correct orientation. The Orbital Life Extension Vehicle (CX-OLEV™), looking like a flying doughnut with solar panel appendages, will be launched from an Ariane 5 rocket. Once in orbit it will use solar ion engines to slowly drift up to the satellite, taking 120-180 days to catch them. Once docked to the satellite, the spacecraft will dock itself to the satellite using a universal attachment. The CX-OLEV will extend the satellite's life by setting it in a snug geostationary path. It will even stay with the satellite to keep its north-south orientation continually adjusted. This technology should save the satellite operators millions of dollars over several years.

For every large deployment, a recovery "sentinel" satellite could be ready to help, This will make satellite builders' jobs a little easier if know they are not totally liable for the expected lifetime of the satellites. The Orbital Life Extension Vehicle can be launched in tandem with healthily satellites to stay "on watch" in case of a system malfunction. This lowers downtime and reduces insurance costs.

In 2004, Orbital Recovery Corporation announced to the American Institute of Aeronautics and Astronautics in a paper by Dennis Wingo, that they counted over 50 working satellites that needed repositioning in high geosynchronous and lower inclined orbits. They stated that there are seven commercial satellites that are in total or partial failure with no chance of recovery without on-orbit assistance. With satellite insurance premiums at historic highs, an on-orbit solution, either from a manned space station or from the ground is needed to mitigate risks. Telerobots like the CX-OLEV, or those now being developed by NASA and the US. Military, may reach an affordability level that insurance companies might help fund.

Building satellites on-orbit is a more progressive application. Satellites have to be tested heavily before deployment. But if engineers are on-orbit, a real-world proving ground could be available for research.

If satellites could be rescued, built, tested and deployed in space, the hardest, most expensive part of the deployment is already over. Large telecom carriers could be relieved of launch costs that surpass the $100 million mark. Rockets are up to 97% efficient, but will still create spectacular failures that bring telecom companies to the very brink. Stage separations are another problem. About 66% of all failed satellites between 1980 and 1999 did not reach orbit because of propulsion problems and bad separations. The Proton, a Russian heavy lift rocket, goes through four stages before reaching high orbit, each one a nail-biter. Just a small reduction in the cost of satellites is in the millions of dollars.

Satellite recovery makes practical financial sense. Insurance premiums are not getting any cheaper as dependency on satellites gets higher than ever. Average costs to insure a satellite is 15% -20% of the total cost, or tens of millions of dollars. Subscribers to radio, mobile communications, and satellite TV will naturally continue to grow with price and availability. Military operations are so reliant on GPS and imaging satellites, that they have developed their own rescue satellite project (Orbital Express) and successfully tested it in space. The problem will soon reach critical levels for commercial carriers if over-reliance collides with insurance rates.

Only with a manned space station can the full cycle of satellite refurbishment be economically exploited. A lucrative market worth millions of unclaimed dollars awaits a crewed effort to re-deploy this otherwise forfeited equipment.

10

Orbital Fuel Depots:
The Big Boost Needed for High Orbits

N ASA envisions a space station that is built simply to refuel spaceships. It will be in the same general orbit as the International Space Station, and will refuel rockets on their way to high Earth orbit or the Moon. It will be commercially funded and maintained to provide a pit stop for vehicles to "top off the tank" on their way to the Moon, Mars, or privately owned space stations. The market would mostly reside in refueling landers leaving from the space stations to Earth, and refueling rockets for their final stages into high Earth orbit to deliver satellites.

Ideas for fuel depots in orbit have been around for years, but were never in NASA's plans. Even today, the new Ares rockets will not be refueled in space, but that does not mean that refueling depots will not be needed in the near future.

At first glance, topping off your tank in space seems like a great idea. The benefits are:

- Economy. You do not have to launch fuel into space. Currently 75%-90% of the weight of a rocket is the fuel. Heavier payloads can be launched with lighter rockets.

- Redundancy. Spaceships can be made reusable. Instead of one-trip expendable rockets, a spacecraft could be used over and over again. Each time they are refueled before they come down or travel out into deep space.

- Safety. Having fuel in space may save lives, due to emergencies in which a vehicle is stranded. Other vehicles in the vicinity can be used for missions they were not intended for, such as emergency landings.

- Going to the Moon. Instead of using an Aries 5, a smaller ship can be topped off and get the Moon in two steps instead of one.

- Simplifying Rocket Design. Staging is a big hazard during the launch of satellites or manned vehicles. Many mishaps happen during the separation process. Vehicles can be designed to have less stage segments.

The problem is that no vehicle exists today that can be refueled in space. You will have to build a ship that can accommodate a refueling process while in orbit. At the same time, you cannot justify that unless you already have a fuel depot up there. What comes first, the chicken or the egg?

The best way to tackle this problem is to have NASA create design changes on their vehicles that can use commercially financed fuel depots. In 2007, one of NASA's new Centennial Challenges for the business world will be to create designs for a fuel depot. According to the preliminary requirements, the tanks must hold at least 20 kilograms of liquid hydrogen and 120 kilograms of liquid oxygen for 120 days. The deadline is 2012 and the reward is $5 million.

If successful, a full scale production model could be scaled up to help both NASA and its industrial partner reduce the cost of business in space. But in the meantime, there is still no spaceship in the fleet that has a gas cap!

Though simplistic on the surface, fuel depots present technical challenges that are fairly steep. One is the type of material used to make fuel storage tanks. They have to handle the stress of the cold liquid hydrogen (-350 F.) and oxygen (-274 F) and the stress of launch. Once in orbit they must handle the sudden temperature changes from day and night (approx: 245 F. to -250 F.) and the pounding of meteorites. This has been done for years, but now the tanks have to stay in orbit for an extended period of time and not leak or break. Heat shields are the best option. NASA has proposed that the tanks themselves should be made of entwined carbon composite fibers. Once baked to a hard cure, these resins are lightweight and chemically resistant.

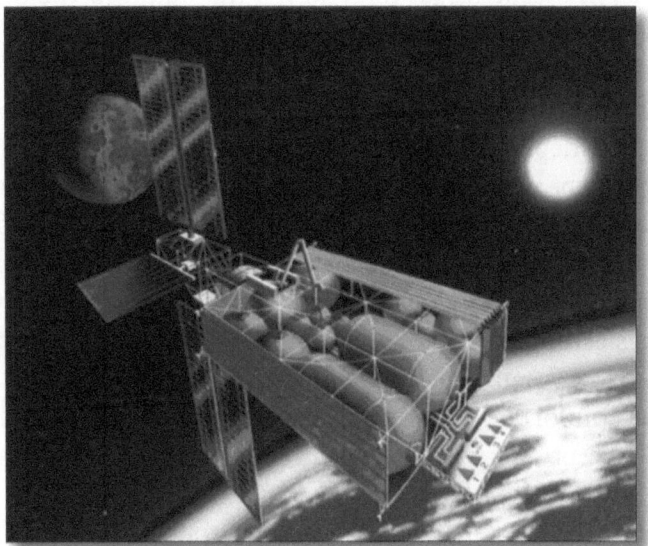

Fuel storage in low earth orbit can help reduce the cost of launches, landings and other operations. *(Image courtesy of Joe Howell/NASA)*

Refueling is not as simple as it looks. The content of the fuel tank will be weightless and not concentrated at the bottom of the tank. This will make it harder to pump out. The fuel needs to be squeezed to one side of the tank so that the pump and conduit can delivery it to the target ship. There also has to be a "gas gage" to measure the level of contents.

The Russians were the first to refuel a space station on-orbit. In the late 1970s and early 1980s, Progress cargo freighters refueled the Salyut 6. The fuel, called unsymmetrical dimethyl hydrazine was fed into the space station automatically. The cargo vessel docked with the station and used pressured nitrogen to push the fuel into Salyut tanks. Like many other operations in the Russian space program, this was remotely automated so that precise control over the dangerous process was possible from the ground. The procedure took two days to complete.

Lockheed Martin Corporation has proposed a simpler method of refueling in space without nitrogen called *Settled Cryotransfer*. After the empty vessel attaches itself to the fuel depot tank, the tank accelerates slowly so that the liquid inside will be pushed to the back end. A pump can then move the fuel from the depot tank to the vessel. The centrifugal force keeps the liquid in the tank from sloshing around. When the vessel is full, it releases itself from the empty tank.

There are predictable commercial benefits from owning an on-obit fuel depot. If NASA will pay to refuel their Moon ships, money could be made from the government. Landings are even more lucrative. A refueled vehicle can land more safely and more accurately that an uncontrolled one. It is also easier on the tourists, scientists, entrepreneurs as well as the delicate products or test results they carry. It is estimated that 58% of a rocket's fuel capacity is used just to get into low Earth orbit. Refueling before returning to Earth would allow for comfortable landings, instead of the bumpy touchdowns, spashdowns or "dead stick" landings previous astronauts had to endure.

Satellite deployment would be another application for fuel depots. It would save satellite companies money by allowing them to use much lighter weight rockets. The vessels need only make to it to low Earth orbit. After refueling, they can go anywhere in the solar system, whether it be an asteroid, or into high Earth orbit to deploy a telecommunications satellite. They never have to be staged if depots are there to refuel them, which means lower insurance premiums because stage separation is one of the most dangerous junctures in any flight.

NASA's Administrator Michael Griffin addressed the American Aeronautical Society in 2005 with an encouraging if generalized suggestion for buying fuel from commercial companies. Though not yet part of the trans-lunar architecture NASA plans to pursue, it could be used for space station re-supply.

As he stated in his address:

> "By the time we are ready to return to the Moon, the
> ISS will have been completed and will be in receipt
> of routine commercial resupply and crew rotation
> service for, we hope, several years. So, if the plan for
> stimulating the development of ISS commercial crew
> rotation capability is successful, it becomes possible to
> envision the crew launch phase of the lunar mission
> being carried out on commercial systems. This would
> be a service we could purchase commercially, leaving
> the very heavy lift requirements to the government
> system, for which it is less likely that there will be
> other commercial applications during this period."

Griffin also went on the say that refueling for lunar flights would
be the next phase in encouraging commercial firms to get involved,
if solely at their own risk:

> "If a commercial provider can supply fuel at a lower
> cost, both the government and the contractor will
> benefit. This is a non-trivial market, and it will only
> grow as we continue to fly."

With the war in Iraq and the International Space Station sucking
up funding, there will be the inevitable budget shortfalls that will
make fuel depots even more attractive in a few years. If history
repeats, it is only a matter of time before the New Vision for Space
Exploration falls into financial trouble and needs commercial carriers
to bail it out. A fuel depot might be the answer for two markets:
supplying both government projects and commercial habitats in low
Earth orbit to keep the space renaissance going.

11

CREW AND CARGO TRANSPORTATION: COMMUTING TO THE COSMOS

S PACE VEHICLES ARE DEVELOPING MORE rapidly than ever before. Trying to keep abreast of developments is impossible, and commercial carriers often cloak their efforts in secrecy until the actual hardware is ready. A few suborbital companies, Scaled Composites, and rocket engine producer, SpaceDev, combined their efforts and showed they can fly to high altitudes, but commercial manned orbital spacecraft are still very much on the technical frontier. Soyuz can be utilized for tourism only if it is also sending cosmonauts to the ISS. Developing a commercial orbital launch vehicle has always been key to opening up industrial markets in space for the common citizen.

There are, however, companies that have the funding and the talent to make it happen. They have been given incentive money from the government to develop manned and unmanned carriers that can reach the International Space Station.

In the summer of 2006, NASA awarded a sum of $500 million dollars to the companies who could best meet the requirements for future space station re-supply needs. Called the Commercial Orbital Transportation Services (COTS) Award, the two companies that prevailed were SpaceX, developers of the Falcon vehicles, and

RocketplaneKistler, developer of the K-1 rocket. The reasons for their successes were that they had real hardware, reusable components and an overall capacity to complete the vision that NASA has had over the years to make reliable and inexpensive re-supply ships.

One of the criticisms NASA has had over the years is that the Shuttle is not only bulking and complicated, but it necessarily has to be manned. The Russians had demonstrated that manned re-supply vehicles are not always necessary or desirable. The Shuttle-like, Buran space plane, first flown in 1988, was a completely reusable spacecraft, launched by an expendable booster that could be flown with or without a pilot, even in bad weather. It had no engines to control, and simply glided down to the snowy steppes of Kazakhstan when its mission was over. The Shuttle, by comparison, requires a pilot. When launching only cargo, this is a liability. Safety requirements resulted in conservative launch schedules and excessive downtime. (Even if there are no

The NASA funded, *Dragon* spacecraft will deploy both crew and cargo to the ISS. *(Image courtesy of SpaceX, Corp.)*

more accidents, the Space Shuttle will have a downtime of 18 percent of its lifetime). Ideally, the launch vehicle should be capable of manned or unmanned flight.

The Russians have two transfer vehicles to the ISS: Progress and Soyuz. The Soyuz is both a launch and return vehicle for human cargo, so it has much more flexibility and economy than the Shuttle when a few thousand pounds are concerned. The Progress spacecraft is designed to be an unmanned re-supply vessel that flies once and is then disposed of. Both have small payload capacity compared to the Space Shuttle, so Roscosmos uses a heavy booster, the Proton, to deploy the large segments of the ISS.

The Shuttle can do more in space than any booster. By comparison, the Shuttle can launch 55,000 lbs into low earth orbit, carry second stage rockets with satellites aboard, can act as an orbital space station for 16 days, and bring down a 41,000 lb payload while landing gently on a runway with seven astronauts aboard. No vehicle can come close to this capacity, but for the next generation of space vehicles, less is more.

The COTS winners will have a more modest requirement than the Shuttle. According to the final announcement made by NASA in January 2006, the spacecraft must:

- Implement U.S. Space Exploration policy with an investment to stimulate commercial enterprises in space.

- Facilitate U.S. private industry demonstration of cargo and crew space transportation capabilities with the goal of achieving reliable, cost effective access to low-Earth orbit.

- Create a market environment in which commercial space transportation services are available to Government and private sector customers.

The challenge is daunting. The next commercial carrier has to build a re-usable system that will not cost NASA the same amount of money normally paid to prime contractors. This "low-cost-high-payoff" model depends on a perfect safety record, steady customers, and a way to define the marketplace. The carrier must also compete with existing government funded supply vehicles. The good part is that the government covers indemnification.

For NASA to buy services, the spacecraft must have the capability to bring cargo delivery to the ISS and dispose of it into Earth's atmosphere when necessary. It also must deliver payloads down to Earth's surface as gently as possible. Crew transportation is required as well, so the spacecraft must be comfortably pressurized, and have a way for the crew to bail out it necessary. These requirements send chills down the spines of those who really know about manned space flight. Nevertheless, the SpaceX's Falcon and RocketPlaneKistler's K-1 spacecraft, will try to meet these challenges and provide launch services for industrial and military markets along the way.

Performance levels are expected to be perfect. For unpressurized cargo, the supply ships must deliver up to 5000 kg (11,000 lbs) per year, and dispose of same amount. Robots and astronauts must be able to off-load and on-load materials, and it must have a reliability of 95% at a 50% confidence level. For pressurized cargo, the supply vessel must deliver 7000kg (15,400 lbs) of cargo, and must be able to dispose of 7000 kg (15,400 lbs) of waste per year. The vessel must safely return up to 3000 kg of cargo per year to Earth.

Crews need a habitat that will allow safe and comfortable flights to the ISS that include all the standards established over 50 years of flight. The vessel must maintain three astronauts who can operate the vehicle every 120-180 days to rotate crews, with a minimum of two to four flights per year. Piloted emergency landing capabilities are also required in case of an evacuation. The cost per flight must be within the budgets of domestic and foreign investors who will want to use the cheapest vehicles they can buy.

NASA will buy the flights, but there are many providers to compete with and all are government subsidized. In fact, there may be more providers than buyers.

SpaceX plans to launch a powerful, two-stage rocket, the Falcon 9, tipped with a pressurized capsule called the *Dragon*. The launcher can also be used to compete with existing satellites launchers with a significant price reduction. The competitive edge of the Falcons is that they use only two stages instead of three or four, and have multiple small engines that can compensate should one or more shut down during flight.

So far, tests have shown that rocket science is as difficult as people say. In a recent announcement, Elon Musk, the founder of SpaceX, equated it to building a software program is separate sections, and then expecting them all to work perfectly the first time you put them together. The prototype, Falcon 1, a small version of the Falcon 9, is being tested at the Kwajelein Missile Range in the Marshall Islands (South Pacific). At the time of this writing, one test flight has failed and one has been partially successful.

Proposed Fleet of Orbital Vehicles for
ISS Launch and Supply Services

Vessel (Pressurized)	Source	Country of Origin	Vehicle	Capacity to Orbit	Return Capacity
Dragon*	SpaceX	USA	Falcon 9*	2500-3100 kg or 7 Passengers	2500-3100 kg or 7 passengers
K1-OV*	Rocketplane Kister	USA	K1*	2775 kg or 5 passengers	2775 kg or 5 passengers
Orion*	NASA	USA	Aries 1	3 –6 passengers and up to 400 kg of cargo or 3500 kg	3-6 passengers and up to 400 kg
ATV Jules Verne	ESA	Europe	Ariane 5	7500 kg	6500 kg**
HTV	JAXA	Japan	H-II	6000 kg	6000 kg**
Soyuz	Roscosmos/ RSC Energia	Russia	Soyuz	100 kg 3 Passengers	50 kg 3 passengers
Progress	Roscosmos/ RSC Energia	Russia	Soyuz	1800 kg	1000 – 1600 kg**
Clipper*	Roscosmos/ RSC Energia	Russia	Soyuz 2,3	500 kg with 6 passengers	500 kg with 6 passengers

*Reusable
**Does not land on Earth but burns up in atmosphere when mission is completed

The Clipper spaceplane, proposed by RSC Energia, is seriously considered by the Russian Space Agency, Roscosmos, as a future orbital supply ship. It is simply a powered glider that will be launched from an upgraded Soyuz rocket. It will be able to dock with the ISS and act at a lifeboat in case of an emergency. The spaceplane can stay docked at the ISS for up to 360 days and act as a habitat of sorts. It intends to have the capability 60 flights in its 15-year lifetime. The intended markets are Russian government payloads, foreign astronauts, and private citizens. Tourists can occupy four of the six seats and will pay for the entire flight at $20 million per seat.

The new vessels from NASA, ESA and JAXA are already in advanced stages of development. NASA's Orion space capsule will be launched with the Aries 1 vehicle and will return to Earth using a parachute. The first flight may be by 2015. The ESA's "Jules Verne" and JAXA's HTV will be disposable, one-way supply ships. They have no capacity to return to Earth and will be only used to disposed of waste by returning to the atmosphere to disintegrate

RocketPlaneKistler is developing its K1- OV re-supply ship as a manned microgravity laboratory as well. Launched on the top of the reusable K1 rocket, the capsule provides a pressurized habitat for about 22 hours. Alloy melts, composite castings, or material exposure tests can be performed in this short period of residence time. It will return to Earth by parachute for a ground landing. Commercial and government customers can "hitch a ride" on each other's flights, thereby mitigating costs.

The ISS alone will not have enough people onboard to handle all the capacity. At the most, six crewmembers will be allowed to work on the ISS provided there are life craft to evacuate them. There needs to be another space station on-orbit to absorb the services.

The low cost launch market is very expensive and competitive. Unless the payloads are very lucrative, the services will be stymied by accidents, insurance costs, overhead, and established competition. A new market must be invented to handle the fleet of vehicles. Only a sustained commercial space facility can do that, otherwise you will have launch solutions before you have on-orbit problems.

Hopefully, COTS will create a trust level with the government so that NASA can use commercial on-orbit services in the future. Today, there is a glut in launch vehicles for satellites. Usually, only one quarter of all rocket launches are commercially driven. But a healthy balance of federal and private markets can get commercial launch services back into growth mode. An industrial research facility on-orbit will provide the impetus for this growth.

12

LONG DURATION SPACE EXPOSURE FACILITIES: TESTING THE LIMITS OF PRODUCT ENDURANCE

I N APRIL OF 1984, THE Space Shuttle, Challenger deployed a space station of its own. The Long Duration Exposure Facility (LDEF) carried over 10,000 specimens, as well as a multitude of hardware elements. This unmanned "retrievable" simply exposed materials to the ravages of space to determine the effects temperature changes and radiation. The data obtained is an invaluable resource on the internet for anyone to use who wishes to utilize the on-orbit environment.

Over the past forty years, all space faring countries, including the European Union, had sent retrievable modules into space to see how instrumentation, materials and biological systems would function in the extreme environment of near Earth space. To this day, exposure experiments are stationed outside the International Space Station to test materials for their endurance. The coming Japanese segment, KIBO will have its own exposure pallet.

The Long Duration Exposure Facility was the granddaddy of all retrievables. Its size was 29 feet by 14 feet and was covered with panels and compartments along its exterior. It stayed in a 200-300 mile orbit for over five years. (A total of 32,422 orbits.) Its mission was simple.

Long Duration Test Facility (LDEF) 1984-1990. Exposure experiments were conducted to test the endurance of materials in the space environment. *(Photo courtesy of NASA).*

Expose metals, glass, polymers, films, and electrical/mechanical systems used in space to high and low temperatures, corrosive atomic oxygen, space radiation and meteorites. Engineers were also concerned about the combined, synergistic effects of on the materials.

Because of the Challenger accident in January 1986, the LDEF facility was returned much later than intended. The original return date was to be February 1985, but it wasn't until January 1990 that the Shuttle Columbia brought back the stranded vessel. After six years in space, the LDEF, stuffed with experiments, chafed by the winds of time and space, survived to tell its story.

Though lab testing might give insight, the *combined affects* the space environment plays on exposed materials would always be in question. A new exposure facility will be necessary to complement the ISS pallets already underway, especially when the new Moon program begins.

A commercial facility in orbit can provide onboard and outboard exposure platforms that would provide affordable, on-demand services to businesses and government customers alike. NASA, ESA, JAXA, CNSA and Roscosmos may turn out to be the commercial exposure facilities largest customers, because of the volume of testing that will needed to be done in a short period of time. Assuming that history repeats, time and money will no doubt be tightly constrained during the return to the Moon. This makes exposure testing an attractive market to entrepreneurs.

An exposure facility and research laboratory would provide a retrieval platform for on-going experimentation. A simple pallet, extended from a commercial space station, would expose materials for the time periods needed. Material analysis could be done onboard; such as, tensile strength, oxidation levels, and the general molecular degradation of metals, polymers, ceramics, and spacecraft components. These materials can then be modified in the on-orbit lab and exposed again to gradually improve quality. They do not have to be returned to Earth.

Exposure can be spun off to Earth based companies. Commercial products such as solar panels, paints, inks and hard coatings for terrestrial applications always need improvement. Ultra-violet light, which degrades automotive finishes, house paint, roofing materials, patio furniture, etc. is always of concern when outdoor products are being developed The effects of cold on metals, glass and polymers used in trucks, airplanes, boats and construction equipment can be easily tested and improved in the natural endurance laboratory of space.

13

SEMICONDUCTOR MANUFACTURING: MAKING MICROCIRCUITS IN NATURE'S CLEAN ROOM

S PACE STATIONS ARE NATURAL PLACES to make microelectronics. This is because space is a very clean. The vacuum of space is extremely low (-10^8 torr). This is an order of magnitude lower than vacuums used in microcircuit manufacturing today (10^{-7} torr). Dust particles are the great enemies of microcircuits. Millions of dollars per year are spent in simply keeping microelectronics factories free of dust. The vacuum of space can keep out the harmful particles that can damage delicate circuitries. Semiconductors, that are now a part of many of the technologies we use in daily life, can be manufactured in space without the complex equipment used on Earth.

The most widely known semiconductor device ever used in space was the Wake Shield Facility, an autonomous, free flying saucer-shaped retrievable laboratory deployed by the Space Shuttle in 1995-1996. Built by scientists at the University of Houston, it was twelve feet in diameter and completely automated with its own power supply and navigational system. It functioned as a super-clean platform to make very high quality templates for microchips. The members of the Space Vacuum Epitaxy Center at the University of Houston ran the project using proven hardware on a $15 million budget. In 1998, Spacehab Inc. acquired exclusive license to manage and market the Wake Shield Facility for $1 million, plus any future royalties.

 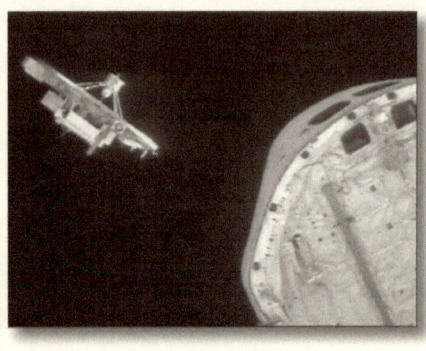

The Wake Shield Facility in the laboratory and outside the Shuttle
(Images courtesy of NASA)

Space is not a complete vacuum. A thin corrosive layer of oxygen still exists even at that high altitude. The Wake Shield Facility provided a reduction in the atomic oxygen normally floating in space allowing the semiconductor crystals to be higher in purity and larger in size than those created on Earth. The facility created gallium arsenide, aluminum gallium arsenide, and indium gallium arsenide films using a gaseous coating technique known as *epitaxy*. These template films would later be used to develop lasers and powerful solar cells.

Automated spacecraft may turn out to be the best method for making semiconductors in space, but should this turn out to be impractical, there is another method under study that might make the process "hands-on" like it is on Earth.

Semiconductors are made in clean rooms. These are places where the air is so clean that only one dust particle per cubic foot is permitted. These facilities are also kept clean from all types of fibers, residues, and even ions. Microcircuit designs are projected on compact disc-sized wafers that are coated with photographic film call photoresist. Hundreds of tiny designs are photographed on the wafers in a checkerboard pattern, each tiny square having the identical microcircuit design. The microchips themselves are later etched onto the wafers inside vacuum chambers, where the air is pumped out to keep contamination at a minimum. In fact, these "wafer fab" facilities actually re-create space environments on Earth.

Clean rooms are very complicated to operate. Millions of dollars go into designing, building, cleaning, maintaining, monitoring, testing, and securing clean rooms. Immense electrical power is used, tons of purified water is used to clean the wafers and work surfaces. Highly filtered air is constantly streaming through the facilities from ceiling to floor. Workers need to wear special "bunny suits", and use clean room compatible tools to keep from introducing contaminants. Robots need to handle all but the most basic processing equipment, because people are too "dirty" to do it themselves. It is an extremely expensive process to make microprocessing chips, yet very high profits are made doing it: $1 million per kilogram in some cases.

To make chips in space, the process must change from a wet to a dry process. Typical photoresist is a chemical that would not survive in space. Fortunately, things actually get simpler. Telerobots do all the work, while receiving signals from the ground. The wafers are not cleaned by water, but by space itself. Ionized oxygen, which is the tenuous "air" in near Earth space, along with the cold and hot vacuum, will scour the wafers in a "dry" wash. There are no liquids used in etching the circuitry on the wafer surface either, because a new process can be applied that doesn't require nearly as many resources as the Earth based "wet" process.

In a study sponsored by the Boeing Corporation, and published by the American Institute of Aeronautics and Astronautics Inc., "Feasibility of Commercial Space-Based Microchip Fabrication" a team of scientists from Boeing, Simom Fraser University, and Andrews Space and Techology, proposed a way to make semiconductors in near Earth orbit using no liquids.

By using light sensitive aluminum oxide coatings instead of photoresist, the microcircuit design would be etched on the wafers without the use of processing liquids. Also, none of the heating, storing, and application processes will be needed either. The resulting higher purity and higher yields, together with a simpler, faster fabrication process would offset the high cost of launching supplies to the space facility.

When compared to Earth-based manufacturing, the space process has many advantages. In the study, the researches concluded a 37% – 45% increase in production speed could be attained in space with 3.6% less electrical power. Designs for new chips can be sent up electronically, fed into the fabrication process, telerobitically processed, and sent back down to Earth for final testing, packaging and distribution. The process would be completely automated with attendant commercial astronauts in case of a breakdown.

In addition to microprocessors, more basic semiconductors can be made in space. These are solar panels for electricity generation and infrared sensors used in electric eyes and vision enhancing imagers. These not only require high vacuum, but also special alloys that are best blended or layered in microgravity. Future robotic products from cars to energy efficient appliances will need environmentally friendly solar power supplies and optical control systems.

With the increasing demand for more and more semiconductors to power our lives, high yields, better performance using less energy is obviously the desired goal. The naturally airless environment of near Earth orbit can simplify the process and quicken the pace of production. Entire space stations could be dedicated for a business as lucrative as semiconductor fabrication, for either full production or research development.

14

Extracting and Refining
Space Resources:
Prospecting in the High Frontier

S PACE MINING HAS ALWAYS BEEN a tantalizing enterprise. Trillions of dollars in materials are available in space. These include iron, cobalt, nickel, titanium, platinum group metals, etc., not to mention unexpected finds of even more precious materials. Satellites, that are old and decommissioned, also possess valuable materials that can be stripped down and made into other products.

For resource extraction, a space station would be useful in simplifying the flight profiles. The problem with mining is getting to an asteroid or a satellite and bringing back substantial payload to Earth. NASA has never even attempted it. But an orbiting commercial space facility could feasibly house a staging and refining operation. To keep the cost of operation under control, the deep space extraction process has to be completely unmanned. The space station would be a place to unload payloads and refuel spacecraft flying to and from the asteroids or satellites. Earth landings would not be needed for every trip.

Commercial space stations could act as places where the raw materials from space can be refined into salable products, from microelectronics to jewelry. If the products themselves were made in space from start to finish, there value would be enhanced.

The easiest material to retrieve would be orbital debris nobody wants, such as old satellites that are expired, broken or hopelessly obsolete. Hundreds of odd satellites orbit the Earth at low altitudes. They are old weather or imaging satellites that contain various metals that can be melted down and refined into valuable microelectronics, machine parts, jewelry, coins or collectables. Scrap metal for strange works of art, motion picture props, educational displays, space station décor, etc. are profitable. The uses are only limited by the imagination.

Boulder fields on asteroids and hardware aboard dead satellites provide "quarries" of available materials. *(Images courtesy of NASA)*

The cost benefit is that no raw materials need to be launched from Earth to manufacture products used in space. Aluminum and silicon would be chief among the materials that would be salvaged. Once the satellites were stripped, converted and reutilized, they would be packed into a different spacecraft that are used to haul cargo to and from the space station from Earth. No space-suited astronauts would be needed unless there was an emergency. The robotic "miner" spacecraft are reused and would never have to make the trip through the burning atmosphere of Earth.

But there will be a limit to how many old unclaimed satellites can be salvaged, so a source of raw materials would have to be derived somewhere else. These would probably come from the asteroid, 3554 Amun. This object has been designated as the closest, wealthiest asteroid to Earth. Popular belief among prospectors is that it contains

$20 trillion worth metals iron, nickel, cobalt, platinum, osmium, iridium, and palladium. This claim can be supported by professional astronomical observations.

There are many crazy ideas on how to mine asteroids. But students at Penn State University created a practical, cost effective method that requires no human presence beyond the space station. In this scenario, a mother spacecraft "barge" carrying a lander and a large canister would fly to an asteroid. The lander and canister would be deployed to the surface. The lander would wheel about using a powerful laser to chop up small areas of the crust. Since Amun is 99% pure metal, no refining on the spot would be needed. Once the rocks are split up, they are collected by a large scoop and placed into the canister. Powerful electro-magnets would keep the lander anchored to the ground during the scooping operation.

The lander would stay on the surface, spending its valuable lifespan mining the rocks. The canister has its own propulsion mechanism to fly up to the barge parked in orbit. The asteroid has very low gravity, so not much fuel is needed. The barge docks with the canister, and goes back to the Earth orbiting space station to empty the payload and get a new canister. In the meantime, another barge/canister arrives at the asteroid to fill up.

The practical economics of this scenario are obvious. The barge itself could use solar ion electric drives to travel back and forth. These are engines that use little fuel and run on solar energy and are now used in unmanned spacecraft today. The barge would use Earth atmosphere to "aerobrake" its speed as it comes by Earth, much like Mars spacecraft do today. After delivering the payload to the space station, the barge goes back to Amun for another run with a new canister. There would be no fiery re-entries of the canisters into the Earth's atmosphere. The reusable canisters would dock at the space station for unloading and be used on the next run. No parachutes, spashdowns or retrieval missions would be needed.

There are caveats, however. Nuclear reactors will be needed to power the landers because asteroids rotate, leaving solar panels in the dark at certain intervals. The landers would have to be heavy and durable, and would have to last as long the Spirit and Opportunity rovers have while exploring Mars. Communication would be sporadic, so autonomous operation (artificial intelligence) would be necessary.

NASA may be interested in participating in the project. One of NASA's goals after the Moon is explore the asteroid belt. Asteroids are believed to contain primordial material from the solar system's birth. A joint government-civilian project to Amun based on the Clementine model (a NASA/Department of Defense spacecraft that flew by the Moon in 1994) could get the asteroid mining industry started. It would provide science an analysis of a pure metal asteroid. This might be of interest to Congress who would ordinarily be trying to reduce NASA spending, rather than increase it.

Of all the space markets, mining is by far the most lucrative, making all the combined gold rushes on Earth look petty by comparison. Amun crosses the orbit of Venus and comes as close as 18 million miles to Earth at certain intervals. Its material wealth could build a new economy on Earth, bringing fabulous wealth to those who can extract the ores.

15

MATERIAL PROCESSING: MIXING AND MOLDING METALS AND CERAMICS

I N THE COMPETITIVE NEW INDUSTRIAL world where efficiency and economy are stretched to maximum limits, manufactures will need a new breed of very strong, lightweight metals and ceramics. The harder and more durable the metals are, the higher the quality, whether the products are golf clubs, or automobile engines, drill bits, or ball bearings. Light sensitivity, electrical conductivity and chemical resistance are also essential, especially for the next generation of consumer electronic products. New alloys are needed for the maritime and aerospace industries to handle the corrosive effects of sea and air, and the strain of high wind sheer and changing temperatures.

New processing techniques developed in microgravity will be one of the ways we achieve this new level of material processing. In space, due to the lack of sedimentation, molten metals disperse more evenly when blended with additives and other metals. In their molten state, there are no convective currents, separating of one metal from another. Dopants, or additives to adjust electrical properties, are better dispersed as well. Microgravity provides the New Physics chemists need to build alloys many times more powerful and more durable than any created on Earth using the same methods.

Metals are blended in vessels or crucibles. One unique benefit of processing metals and ceramics in space is the ability to levitate the molten sample so that it will not touch the side of the crucible. The molten products will not become contaminated by the crucible walls as they do on Earth. This is called electrostatic levitation: the suspending of samples in a magnetic field so that they will not drift and bump into the sides of the container.

This process can be done on Earth, but it requires a lot of energy. Usually only small objects are levitated. But in space, large industrial blending runs could be made using less energy and yielding superior results.

There are many objectives to material processing in space. Current R&D efforts endeavor to:

1. Make the metal and glasses stronger from the outset.
2. Blend alloys and additives more thoroughly.
3. Make very pure and chemically resistant metals.
4. Make perfectly spherical metal products.
5. Making metal foams and ceramics in a flawless way.

The products created are for:

1. High efficiency solar panels
2. Microprocessing semiconductors (for computer chips)
3. Powerful magnets for industrial tools
4. Electric eyes, night vision instruments, and medical optical devices.
5. Human body prostheses

Stronger Microstructures

Metals and ceramics made in space are pure and free of flaws. This purification can enhance their physical properties. Steel, for instance, is made from iron that has been purged of most of its carbon impurities. Iron by itself is brittle and subject to rust. Steel is stronger and more flexible, and when mixed with chromium, becomes stainless steel. If purity efforts are taken a step further, metals can be made that are amorphous (non crystalline). By controlling this level of crystallization, metal can be made more elastic or "springy", thereby increasing their durability.

The microstructures of metals, called dendrites, are treelike molecular chains that dictate how metals behave. Chemist can improve the quality of the metal by changing the dendrite structure. On Earth the dendrites will grow in a distorted pattern due to the force of gravity pulling the microstructures in predictable directions. By growing metals in microgravity, dendrites form in their natural crystalline states due to the fact that there is no convection or settling in the melting and solidifying process. Pure metals made in the vacuum of space, allow for the development of specialty products that are used in critical applications where breakage could result in serious losses of human life. Medical devices, building materials, and highly reliable aircraft engine parts are possible uses for these metals.

Metallic Glass (Amorphous Metals)

During space processing, purity levels can be so high, that the metals take on the smooth qualities of glass. This is called homogenous nucleation. If there are no impurities in the blends, the metals and alloys can be under-cooled below their normal solidifying points. Once at the critical temperature, the molten alloy forms microstructures that are amorphous. Unlike window glass, amorphous metal glass is stronger. Diamond is an example of graphite under going homogenous nucleation. Its sister compound, graphite, is the same material, but in a heterogeneous nucleation state.

Glass is much easer to process than metal. Because glass has a lower melting point, it is easier to mold into desired shapes that metal. Machine parts can be cheaply and quickly manufactured if they are made of amorphous metals. Less labor-intensive milling or machining is needed. The processes must allow the complete purity of the melts, and undisturbed solidification of the final blends. Metallic glass is so durable, that even steel ball bearings can bounce off it. Microgravity facilities in the vacuum or space can provide a clean room laboratory for process development so that the alloys made there are perfectly pure.

On particular alloy of interest is (Zirconium-Nickel-Copper-Titanium-Beryllium). This product, if processed in space, could be molded easily into a desired shape, would not shrink after it cools, and would not require finishing. Normally a tremendous amount of heat is needed to forge metals into these desired states, but with a containerless facility in space, large batches could be made more easily than on Earth. The aerospace industry is very interested in making airframes and engine parts out of amorphous metals. Once combined with other stabilizing materials, they are lighter and superior in strength to aluminum.

The Quest for Super-Spherical Ball Bearings

Ball bearings make our industrial civilization operate. Better types are always needed. Without them, cars, trains, assembly lines, etc., will grind (literally) to a screeching halt.

Making ball bearings in space is easy. In a liquid state, metals and alloys will take on a round shape due to their surface tension. If scientists could figure out how to make ball bearings on Earth that were perfectly round, the discovery could lower the frictional wear and tear on any machinery using ball bearings today. These would help with energy efficiencies in an age where every ounce of power is extracted from machinery for the least cost. Space processing may give chemists ideas on how to make better bearings them on Earth.

Can ball bearings be produced in large capacities on-orbit? There are a few challenges in making ball bearings space. First, will customers pay for better quality? Second, is security. If ball bearing manufacturing were used for military hardware, the space station would become a legitimate military target for enemy attacks! Mundane manufacturing may take longer to develop, but free transportation, to and from the space station, would be a boost to suppliers.

Examples of Commercially Relevant Metals and Alloys Made in Earth Orbit

Mission	Blend	Product Uses
Skylab	Gallium Arsenide Indium Antimonide	Solar panels, microprocessors.
	6mm metal spheres	Ball bearings
Shuttle Spacelab MSL-1 (ESA –Germany)	Zirconium-Copper-Aluminum-Nickel Zirconium-Copper-Aluminum-Cobalt Zirconium-Copper-Titanium-Nickel Zirconium-Copper-Aluminum-Nickel- Niobium Iron-Chromium-Nickel Cobalt-Palladium Palladium-silicon Palladium-Copper- Silicon	Undercooled glass for metal replacements.
	Lead -Tin-Telluride Indium Antimony Aluminum-Lead-Bismuth	Electro-optical Devices. Superconductor
MIR	Zinc Oxide and Yttrium-Barium-Copper oxide Gallium Arsenide silver/germanium, antimony, lead-silver chlorides, Aluminum/nickel, gallium, gallium/ antimony and aluminum/copper/iron.	Superconductor Solar panels and semiconductors.

Space Foams

While preparing metals for space processing, gas can be added to the molten blends that will make the metals bubble. These bubbles will not reach the surface and pop like they do on Earth, but will stay as holes in the metal products. This is because the gas bubble weighs exactly the same and the metal. The gas bubble does not rise to the surface. Normally this is not desirable, but these vascular (hole-filled) metals could be used as substitutes for solid materials because they are more lightweight and just as strong.

These "space foams" are an object of ESA research aboard the ISS in the coming years. If successful, stronger and more lightweight metals and glasses can be developed to improve land, sea and air vehicles. Because weight is a big cost factor in transportation, all types of engine and body parts for jets, automobiles, and spacecraft could be made with half the weight, and just as strong. The holes will not weaken the strength of the metals. Cars can be built that are light and tough, so that hydrogen fuel cells can run them. When accidents occur, less damage and loss of life in collisions would occur. Metallic foams also do not conduct the heat and cold as much as solid metals, and they even have better sound proofing properties as well.

Foaming processes are not exclusive to super alloys. Biocompatible ceramics, foamed in space can create bone transplants for injury victims because the porous nature of the prosthetic material more mimics that of real bones. Human bones are porous yet strong. To best patch them up, a porous, yet extremely tough implant material is needed. Porosity allows for better healing and longer life. Research on foamed ceramic bone replacements have already been conducted using calcium and phosphate compounds by Dr. Frank Schowengerdt and Dr. David Moore working at the Colorado School of Mines. Using an experimental furnace aboard the KC-135 "vomet comet" aircraft, the scientists found that escaping gases from the molten ceramics created the desirable bone-like pores. But there is still a long way to go before a product is born.

Bone replacements are big business. This promising field holds staggering market potential. Implants now last about ten years and then have to be replaced through painful surgery. Aging Baby Boomers will need boatloads of hip and knee replacements in the coming decades, not to mention the future war veterans who could greatly improve their lives with longer lasting implants. Because of the inevitabilities of injury and old age, the number of implants will be needed in the U.S. alone should be staggering.

TEMPUS

Many experiments on the Shuttle and MIR space stations were simple. The alloys were heated up until they became a gas. They were then transported down a tube, and then re-crystallized in a much more purified form. The mixing and solidifying or the gas in microgravity yielded a more pure and evenly blended alloy microstructure.

Another method, mentioned earlier, involves levitating the sample in an alloy furnace. This is actually operational today. It is called the Electromagnetic Containerless Processing Facility (TEMPUS), a joint German-American project first conducted in the Space Shuttle. In space, the sample floats in the air and is kept from touching the container walls by using magnetic coils to keep it from drifting. The furnace can go to temperatures of 2500 C. to thoroughly mix the compound. The temperature can then be dropped until the desired properties are attained. Because the sample does not hit the sides of the container, it remains free of impurities that would affect its structure. The result is a perfectly pure metal or alloy with glass-like properties.

Space Shuttle astronauts proved containerless processing was feasible aboard the International Microgravity Science Lab-1in 1994. On a subsequent flight in 1997, TEMPUS hosted 22 experiments comprising 197 hours of test runs and 437 melting cycles. It successor, the Advanced Electromagnetic Levitation Facility, was supposed to be an addition to the ISS, but instead had to be performed on Airbus A300 Zero G Parabolic Flights by the German Aerospace Agency.

During the Shuttle era, Japanese engineers developed the Large Isothermal Furnace. Melts included lead-tin-telluride, and blends of silicon, gallium, germanium and antimony. The Japanese Space Agency (JAXA) has taken a particular interest in alloy research in space, and this field of research will be one of the focuses on their KIBO module for the ISS. The Electrostatic Levitation Furnace (ELF) will be the facility JAXA uses in the ISS to boost the Japan's already dominant role in the industry. When the ISS is decommissioned, a commercial facility must be up and running to continue the research.

TEMPUS, a device for studying metals in space, controls the motion of levitated samples (about the size of a gumball) in electric coils at temperatures of 2500 C. *(Image Courtesy of NASA)*

Material science touches every area of our lives, whether it is at work or at play. Space research techniques can improve ground based manufacturing. The low energy requirements of on-orbit levitation, makes manufacturing attractive, especially for critical components used in medicine and transportation. As energy efficiency standards become more and more stringent, there will be a bright future in creating products that meet these new challenges.

16

VISIONARY MARKETS:
MENLO PARK IN ORBIT

THERE ARE MANY EXCITING IDEAS for making money in space. So many that I cannot mention them all here, but one thing is for certain: there is a need for experimentation. It is a market unto its own: the need to tinker and the need to quickly test an idea. Governments and large businesses of the world would be delighted with an affordable way to utilize space for quick tests. Small businesses with ideas for killer applications may need only one test. Given the expense of using it ISS, the commercial space station is the only other alternative for comprehensive research for entrepreneurs.

Visionary microgravity products may be in the very early stages of development, or completely hypothetical. The products may be a part of an even bigger project created on the ground, perhaps a dress rehearsal for use in the exploration of the Moon and Mars. They would not pay for themselves right away, and maybe not produce profits at all. But down the line, cleaner energy devices, better household appliances, etc. could "spin off" into a large profit. A case in point would be cordless electric drills that are now common in the marketplace. These lightweight reliable products are a spin-off from the Apollo program. Historical films of astronauts drilling on the Moon were the first demonstrations of this technology to the public.

In an Edisonian laboratory in space, old technologies would be improved and brand new ones discovered. The breakthroughs unleashed could rival or surpass those of the Industrial Revolution. The paying customers would cover a wide spectrum of science and entertainment projects. Some of these will be experiences captured on film, or concepts stored in databases, such as:

1. Government science and exploration contracts to support their own technical and material competitiveness.

2. Companies that want to perform raw (one-time) experiments to verify feasibility.

3. University payloads for students.

4. Entertainment film studios for documentaries, thrills, arts, games, and mainstream features.

5. Clean energy technologies that could reverse global warming and conserve energy.

Since so much of this is unclear these days, how can investors be sure research stations in space will be profitable? The answer is that the world governments will be the first customers! After the International Space Station is decommissioned, the space agencies will have no choice but to hire a commercial facility to test flight hardware.

NASA is already taken an interest in the first baby steps of commercial space. Over the past few years, NASA engineers have made repeated trips to Las Vegas to see Robert Bigelow's inflatable habitat projects. NASA is eager to use these types of inflatable habitats on the Moon. In 2007, NASA drafted a memorandum with Virgin Galactic, LLC to provide services for developing spacesuits, heat shields, and hybrid rockets. In the same year NASA also contracted the Zero Gravity Corporation for testing hardware. In June of 2005, DARPA, the R&D department of the U.S. military, used the Scaled

Composite's White Knight booster plane, used with SpaceShipOne, to test a military space plane, the X-37. In other words, NASA and DARPA will contract services with private companies as a cost saving measure. Congress will soon not let them do otherwise due to budget oversight pressures.

But why don't they use the ISS? Because NASA does not have the funding, time or the personnel available to perform the tests! As long as there is more distant goals for NASA to pursue, such as the Moon or Mars, the space station work is better contracted out. This is especially important for other governments around the world who never could reach space at all. If the lease rates are low enough, the government could share expenses with large commercial companies, instead of wrestling with Congress for extra funding.

In all my years of working in research and development laboratories, feasibility studies were one of the most important areas. Feasibility included tests to improve quality, demonstrate unique concepts, and scale up prototypes into pilot production runs. If you have enough time in space to tinker, R&D can turn into process development and even to full on-orbit production. The products are myriad.

Colloids are solids suspended in liquids. They are used for chemical coatings, and food additives. Coatings are used in painting, printing, finishing, corrosion inhibiting, antioxidizing (sunlight protecting), light absorption, weatherproofing, and abrasion resistance. When these products are made on Earth, gravity distorts the mixing of chemical ingredients through sedimentation. Novel blending in a microgravity setting can improve quality. NASA has been tinkering with colloids in the ISS to find ways to making better optical displays. New generations liquid crystal displays would depend on industrial chemists turning these novel chemistries into salable products.

Polymers (plastics) are used in nearly every consumer product available. The textile industry weaves polymer strands into the clothing we wear and the carpets in our homes. In the manufacturing stage, mixing is very important for polymer reactivity. It dictates the density and molding characteristics of the polymer. Without good mixing, the product is weak. Novel blending in space may improve polymerization. It also would allow chemists to make new types of stronger and more flexible polymers for the future where energy related warfare and environmental conservation are playing hand in hand.

Custom processing in space works with most products, and niche marketing is the key. Small space markets can create specialty products that look and function better and have the appeal of being space made. Products like time-release drugs, coated and blended in space, or space gems such as ruby, emerald, sapphire and opal, synthesized on-orbit, and fiber optics used in medical lasers, are examples of niche markets that would have high profits to match. Will niche products be profitable from space? Only experimentation can tell.

Energy is the most urgent crisis of our time: both in its generation and in its waste management. Shuttle, MIR and ISS studies have been going on for years examining ways of developing all kind of solutions. Growing large zeolite crystals in space in one. These sponge-like rocks made of silicon and alumina can assist in refining oil and could have a dramatic effect on stretching domestic supplies. Early ISS experiments indicated long periods of growth in microgravity increases the sizes of zeolites by 175%. These stones may be engineered to contain gaseous hydrogen so that it can be stored and transported without the use of heavy tanks. Only an industrial research scientist can truly develop these stones for a commercial profit.

Combustion research has also been intensely studied by astronauts and cosmonauts. In space, flames burn as a small blue ball of light without any soot coming out: a perfect sphere of theoretical efficiency. But how? How can we duplicate this perfection on Earth where 85% of our energy comes from some form of combusted fuels? Think of the profits generated by pollution free smoke stacks? Can combustion studies be completed by the time the ISS is decommissioned? A solution to soot production could save energy companies millions.

Agriculture has been a field of steady research in the Salyut and MIR long duration flights. It is a difficult endeavor in space, but many problems have already been overcome by persistent Soviet research.

Space commercial agriculture comes in three principle forms:

1. Seed breeding

2. Lignin reduction
3. Greenhouse development for oxygen production and food.

Seed breeding is a popular discipline with Chinese space program. By sending crop seeds into space aboard a retrievable capsule, the seeds experience genetic mutations due to the cosmic radiation. According to the Chinese space agency, CSNA, higher crop yields occur form seeds exposed to the space environment Since the seed breeding program began in 1987, crop yields have risen by 10% -20% for exposed seeds. It is well known that irradiating seeds can mutate genetics in plants, but what role does weightlessness play?

Lignin is the chemical in plants that make them rigid. This "stiffness" allows plants to stand up to wind, snow and rain. But when you're trying to make paper, lignin resists cutting. Lignin is also not digestible in humans. Not only can low lignin plants be used for cheaper paper, but can be used as a food additives to stretching existing supplies. Food additives and genetically altered crop seeds are doubtless profitable enterprises that have short-term returns on investments. Only refinement on-orbit can see these products reach their full potential.

Space greenhouses take many years to perfect. Only recently did astronauts find out that good ventilation helps plants reproduce properly on-orbit. The Russians tinkered for years in Salyut and MIR to find a way to grow plants for oxygen and food production. Nothing discovered so far can come close to satisfying the need for space settlers. If plants are susceptible to disease, insect infestation, and fouled water supplies, how can they be depended on for something as crucial as oxygen? According to estimates, astronauts need 30 square meters of plants per person for food, and 10 square meters of plants per person for air.

Existing space hardware needs to be redesigned, particularly for space suits. Space walking is a complicated and dangerous occupation that relies on perfect performance of the suit. The insulating efficiency and material flexibility of spacesuits could mean the difference in how long an astronaut can function. New materials are being engineered to improve mobility and radiation protection in spacesuits. Comfort is crucial for long term walks in space or on the Moon, and the ability for the suit to keep an astronaut warm and cool is always desirable. As with most products, cost reduction is necessary as well. The best proving ground for effectiveness is in actual flight from commercial facility on-orbit.

Space stations are a great place to rehearse on-orbit tasks. Activities such as space welding, on-orbit repairs, solar panel installation, engine test firings, liquid refueling, radiation shielding tests, artificial gravity machines, astronaut training, would be better held in space instead of in the water tanks used be NASA now. The world space agencies, ESA, NASA, CNSA, JAXA, CNSA and Roscosmos will need affordable space workshops to make the Moon and Mars initiatives financially sustainable. Practical on-orbit training is essential if we are really serious about manned exploration of the solar system.

Science fiction has always been dominated by film because it is such a visual art. From Buck Rodgers to Apollo 13 (the movie), these films have been a 20th century art form that is unique to our times. In Ron Howard's 1995 film, Apollo 13, he proved that space stations could be used as studios. Some of the scenes inside the Apollo 13 vehicles were shot in the parabolic flying airplane, KC-135, also known as the "Vomit Comet". The airplane was used to simulate weightlessness for about 20 seconds to train astronauts. It does this by making extreme dives from a height of 36,000 feet. Howard placed a movie set inside the plane where actors could do stunts and say their lines without an elaborate set. During the filming, eleven actors and crew flew up and down for four weeks. The film grossed $355 million worldwide. But there were many difficulties. The drone of the engines affected the films sound quality, and the plane was pushed so hard that it had to be grounded for suspected cracks in the engines

A space station could become a movie studio where all types of film could be made in a stable environment. Markets include sweeping historical space dramas, educational documentaries, music videos, and adult sex features.

Weightlessness would be a great novelty for the adult film industry. Actors can assume any position imaginable. Blood flows unhindered in microgravity, so it circulates through the sex organs, actually boosting activity. (Shuttle astronauts have reported this) Lush breasts and hair float in "wild thing" fashion, making for titillating poses.

Sets or rehearsals are cheaply made, and the films can be sold on the internet for a small fortune.

A word of caution to future adult film entrepreneurs: as much as the adult film industry is part of our economy, there will be some investors who may not participate in a commercial space facility that uses extramarital sexual activity to subsidize itself. Some call it hypocrisy, but this must be carefully considered in order not to offend big investors or educational institutions that are answerable to parents and teachers. As tempting as space pornography may be, there are dangers in embracing it as microgravity market. We must remember that most of the world is still opposed, at least in principle, to premarital and extramarital sex.

One of the reasons space travel has always been "G" rated is because it is tied to education and young people. If a commercial space station were ever viewed as a "brothel in the sky", NASA and the educational institutions it depends on, may have to refrain from using it due to public and parental pressure. Large corporations, known for their political correctness, may be embarrassed and feel pressured to defend themselves. And if there were a pregnancy on-orbit, the delicate fetus might be damaged by radiation, intensifying the controversy. If this ever does happen, the lucky lady will have to leave space immediately, and there better to be "chariot" ready and waiting!

Babylon will certainly follow us across the stars, just as it did across the continents. But how will it be regulated if on-orbit pregnancy is considered as a high-risk medical condition? It is as much a social and legal frontier as it is technical one.

Mainstream feature films that indulge in sex are not so easy to reproach. The microgravity environment is a wonderful place for romance: the glow of the Earthlight, the brilliance of the stars above, the beautiful moonrises and sunsets that even Venus would admire. Naked bodies floating feely like living erotic art, close encounters of the 4th kind between consenting adults in a flowing dance of lovemaking. Mainstream "R" rated films are more palatable to parents and teachers if well known movie studios are producing them. These studios also have the deep pockets to produce them in space.

Sport is another experimental market. No one knows whether games can be organized into real sports in space. How do you practice? How to you refine your athletic skills? Are there sports in the Antarctica? On Mount Everest? Under the ocean? Sports in space have to be rehearsed on Earth, and athletes have to be very well adapted to space to do all these gymnastics without getting space sick. They will have to use a section of the space station that rotates so that their bones won't dissolve during the years of practice. Soccer and tennis require only simple equipment, and use goals on either end of an open area. Instead of boundary lines, electric eyes alert the players when the ball goes out of bounds. When the ball gets behind a defender enough times...match point.

Personal combat in microgravity would be interesting. One-on-one matches of space boxing with each opponent, knocking each other out by the sheer pummeling and kicking in the dizzying climate of weightlessness, with fake blood festooning out and congealing in clouds of red. Like the in days of ancient Rome, gladiatorial combat may be revived to its former peak in popularity.

Dance takes on the same artfulness as the water shows of old Hollywood Busby Berkley movies. Without the need to come up for air, great performances could go on for hours without the dancers getting tired. Circus Solei at a slow more leisurely pace. The dancers would move more like ghosts or angels, flying through heaven and Hades. Human art forms in perpetual motion, recorded on film from the "Studio in the Stars". A product packaged and sold in real time that could be beamed down onto billboards for all to see. Advertising becomes an irresistible business to exploit with the eye-catching, style, fashion and eroticism inherent in popular space travel.

As commercial space becomes more popular, bohemia will be hitching a ride. As NASA relies more and more on the space commercial sector, artists will find ways of exploiting the High Frontier. Visionary markets are as much a means to self-discovery and expression, as they are to technological and economic expansion. As we grow outward, we will grow inward.

AFTERWARD:
AEROSPACE IN THE FUTURE

S INCE THE END OF THE Cold War in 1990, aerospace employment has declined dramatically. After enjoying a high of 1.1 million jobs in 1990, the aerospace workforce had been cut nearly in half by 2007 to 629,000. Market forecasts by the FAA show that aviation business will increase in the coming years, but that is mostly due to foreign markets catching up to the west. The space aviation market has long since matured to the point of graying, and needs stimulus to stay healthy.

Civil space aviation employment at NASA has leveled off as well. In its heyday in the mid-1960s when the Apollo program was in high gear, there were 35,800 permanent employees (1967) with 374,000 contractors, now NASA employs only 18,600 permanent workers and 43,500 contactors. The workforce has aged considerably since the Cold War. According to NASA's own HR records, in 2007, 78% of its civil employees are 40 and over.

Prospective scientist and engineers do not seek as many degrees in aerospace as they do in electrical or mechanical engineering. Aerospace professions reached a crises level in the early 2000s, when then NASA Administrator Daniel Goldin had stressed that this problem was "overwhelming" and that NASA could not continue with its ambitious plans without many more younger, qualified aerospace workers. This can be traced back to the fact that most students do not believe the space program is a viable career goal or is important to the economy as a whole.

In a study reported at the American Institute for Aeronautics and Astronautics in 2006, Dittmar Associates, a firm that deals in strategic planning and government research, found that people, aged 18-24, had a dimmer view of space exploration than their baby boomer parents and grandparents. Interest in the return to the Moon was only 29%, with 45% neutral, and 23% disinterested. They are much more concerned about jobs, relationships, and war.

The attack on America on September 11th, 2001 is an alarm still sounding in the minds of western civilization. The overarching lesson from that fateful day is that energy extracted from foreign sources cannot be relied upon forever, and that this dependence can be exploited for political gain. Space research and manufacturing will help provide tools to create alternative domestic energy sources and the new technologies that can support them. By making faster computer chips, building strong lightweight metals, and efficient solar cells, we can reverse the bitter inheritance that we have bestowed upon both our children and our Earth.

Commercial space is part our new economy: one that gives us more opportunities for growth. A commercial space station could establish a new world of benefits most of which we cannot possible realize. Under a crown of space habitats, the world economy can take advantage of renewable energy, medical advances, and new travel experiences that will evolve human culture to its next level. Instead of being mired in war, or stalemated in a cold peace, we can solve our problems and change our futures.

Bibliography

Web Sites:

Chapter 1
www.faa.gov
www.nasa.gov
www.panamsat.com
www.intelsat.com
www.marisat.com
www.futron.com
http://www.sia.org
http://history.nasa.gov\
www.satelliteretailers.com

Chapter 2
http://science.nasa.gov
http:/science.nasa.gov
http://history.nasa.gov/SP-4011/app3.htm
http://www.hq.nasa.gov/office/pao/History/SP-4209/toc.htm
http://history.nasa.gov/EP-107/contents.htm
http://ntrs.nasa.gov/archive/nasa/casi.ntrs.nasa.gov/19870012563_1987012563.pdf

http://www.hq.nasa.gov/office/pao/History/SP-4209/ch1-3.htm
http://home.comcast.net/~rusaerog/mir/Mir_exp.html#7
http://www.energia.ru/english/energia/mir/mir-science-06.html
http://history.acusd.edu

Chapter 3
www.spaceislandgroup.com
www.bigelowaerospace.com
www.spacehab.com
www.spacefuture.com
www.alliancespace.net
http://www.nasatech.com/Spinoff/Spinoff2004/hm_5.html
http://aerospacescholars.jsc.nasa.gov/HAS/cirr/ss/2/4.cfm
http://www.astronautix.com/project/mir.htm
http://www.esa.int/esaHS/SEMBCY1PGQD_business_0.html
http://idb.exst.jaxa.jp/english/home_e.html
http://erg.usgs.gov
www.thespaceshow.com

Chapter 5
http://ssi.org/?page_id=25
http://www.spaceref.com/news/viewpr.html?pid=2571
http://www..spaceislandgroup.com
http://www1.eere.energy.gov/solar/pdfs/solar_timeline.pdf

Chapter 6
http://science.nasa.gov/msl1/themes/benefits/natures_pharmacy/
index.htm
http://www.cbse.uab.edu/ongoing-projects.shtml
http://www.cbse.uab.edu
http://www.energia.ru/english/energia/iss/researches/space-bio-
12.html
www.thespaceshow.com

Chapter 7
http://science.nasa.gov/NEWHOME/headlines/msad05oct99_
1.htm
http://www.colorado.edu/engineering/BioServe/about.html.

Chapter 8
www.spaceislandgroup.com
www.spaceadventures.com

Chapter 9
www.orbitalrecovery.com
www.boeing.com
www.nasa.gov

Chapter 10
www.nasa.gov
www.lockheedmartin.com

Chapter 11
www.spacex.com
www.faa.gov
www.rocketplanekistler.com
www.nasa.gov
www.esa.gov
www.astronautix.com
Chapter 12
www.nasa.gov
http://setas-www.larc.nasa.gov/LDEF/index.html

Chapter 13
www.nasa.gov
http://idb.exst.jaxa.jp/english/home_e.html
http://idb.exst.jaxa.jp/db_data/summary/english/e_seika_gaiyou_index.html
http://spaceflight.esa.int/eea/index.cfm?act=search.basic
www.uh.edu
http://www.energia.ru/english/energia/mir/mir-science-06.html

Chapter 14
www.nasa.gov

Chapter 15
http://idb.exst.jaxa.jp/english/home_e.html

http://idb.exst.jaxa.jp/db_data/summary/english/e_seika_gaiyou_index.html
http://spaceflight.esa.int/eea/index.cfm?act=search.basic
www.nasa.gov
http://www.energia.ru/english/energia/mir/mir-science-06.html

Chapter 16
www.spacefuture.com

Afterward:
http://www.dittmar-associates.com/Market%20Study%202006%20Update~web.pdf
www.nasa.gov

Books:

Thorpe, Andrew, *The Commercial Space Age: Conquering Space Through Commerce,* AuthorHouse Publishing, 2003

Linenger, Jerry M., *Off the Planet: Surviving Five Perilous Months Aboard the Space Station MIR*, New York, McGraw Hill, 2000

Compton, David W., Benson, Charles D., *Living and Working in Space: A History of Skylab*, Washington, DC, NASA. 1983

Clark, Phillip, *The Soviet Manned Space Program: An Illustrated History of the Men, the Missions and the Spacecraft*, Orion Books, 1988

Harland, David M., *The Story of the Space Shuttle*, Praxis Publishing, 2004

Dorsey, Gary, *Silicon Sky: How On Small Start-up Went Over the Top to Beat the Big Boys into Satellite Heaven*, Reading Massachusetts, Perseus Books, 1999

Simpson, Theodore R, *The Space Station*, New York, IEEE Press, 1985

Heppenheimer, T.A. Countdown: *A History of Space Flight*, New York, John Wiley & Sons Inc, 1997

Von Braun, Wernher, Ley, Willie, Whipple, Fred L., Kaplan, Joseph, Haber, Heinz, Schachter, Oscar, Ryan Cornelius, *Across the Space Frontier* New York, Viking Press, 1952

Lewis, John S., *Mining the Sky: Untold Riches form the Asteroids, Comets, and Planets,* Reading Massachusetts ,Helix Books, 1997

Schmitt, Harrison H, *Return to The Moon: Exploration, Enterprise ,and Energy in the Human Settlement of Space,* New York, Copernicus Books, Praxis Publishing, 2006

Nansen, Ralph, *Sun Power: The Global Solution for the Coming Energy Crisis*, Lopez Island Washington, Ocean Press, 1995

Kluger, Jeffrey, *The Apollo Adventure: The Making of the Apollo Space Program and The Movie Apollo 13*, New York, Simon and Schuster, Inc., 1995

Borrough, Bryan, *Dragonfly: NASA and Crisis Aboard MIR*, New York, Harper Collins Publishers, 1998

ABOUT THE AUTHOR

Andrew M. Thorpe has spent the last 26 years engineering products and services for private companies. He began his chemistry career developing new polymers and custom blended pharmaceuticals. He then switched to consulting, where he acted as a technical advisor and marketing liaison for the development of microcontamination control solutions for high tech environments. During his years as a consultant, a wide variety of industries were supported including aerospace, biomedicine, semiconductors, optics, and information technology. He then began developing software systems for scientific, financial, legal, and medical management firms to provide better collaboration environments.

As a freelance science writer, he has appeared in Astronomy, Sky and Telescope, Omni, and Boy's Life magazines. His last book, *The Commercial Space Age: Conquering Space Through Commerce*, was published in 2003.